WITHDRAWN

D1207492

America's Ocean Wilderness

America's Ocean Wilderness

A Cultural History of Twentieth-Century Exploration

Gary Kroll

University Press of Kansas

Published by the University Press of Kansas (Lawrence, Kansas 66045), which was organized by the Kansas Board of Regents and is operated and funded by Emporia State University, Fort Hays State University, Kansas State University, Pittsburg State University, the University of Kansas, and Wichita State University

Library of Congress Cataloging-in-Publication Data

Kroll, Gary, 1970-
America's ocean wilderness : a cultural history of twentieth-century exploration / Gary Kroll.
p. cm.
Includes bibliographical references and index.
ISBN 978-0-7006-1567-4 (cloth : alk. paper)
1. Naturalists—United States—Biography—History.
2. Explorers—United States—Biography—History. 3. Discoveries in science—United States—History. 4. Underwater exploration—United States—History. I. Title.
QH26.K76 2008
551.46072′3—dc22 2007045598

British Library Cataloguing-in-Publication Data is available.

Printed in the United States of America
10 9 8 7 6 5 4 3 2 1

The paper used in this publication is recycled and contains 50 percent postconsumer waste. It is acid free and meets the minimum requirements of the American National Standard for Permanence of Paper for Printed Library Materials Z39.48-1992.

Contents

Preface, vii

Introduction, 1

1 The Oceanic Hunting Grounds of Roy Chapman Andrews: Whales as Resource and Game in Postfrontier America, 9

2 Robert Cushman Murphy and the Natural History of Ocean Islands: Conservation and the Myth of the Inexhaustible Ocean Frontier, 37

3 Sensational Management: William Beebe and the Natural History of the Ocean Sublime, 65

4 Rachel Carson's *The Sea Around Us:* The Construction of Oceancentrism, 95

5 Eugenie Clark and Postwar Ocean Ichthyology: Gender, Oceanic Natural History, and the Domestication of the Ocean Frontier, 124

6 Technophobia and Technophilia in the Oceanic Commons: Thor Heyerdahl and Jacques Cousteau during the American Cold War, 152

Conclusion, 189

Notes, 195

Selected Bibliography, 223

Index, 241

Preface

More than a few of my friends think it a bit odd that I left South Florida to go to school in Oklahoma, where I wrote a book about the ocean that was published by the University Press of Kansas. But to truly know the ocean, you simply need to know a bit about the West, the frontier, and the wilderness—or rather, you need to know about these places as ideas and concepts. Before moving to the Great Plains, I came to know the ocean in the same way as my suburban friends and family: by watching TV. Yes, I spent plenty of time at the beach. I spent hundreds of hours swimming and snorkeling, and I gave scuba diving a try as well. But the ocean that I came to truly love and care about was not the little strip of dredged sand at Hollywood Beach. No, I had traveled to the ends of the earth in the *Calypso* with Jacques Cousteau. He visited coral reefs that made John Pennekamp Coral Reef State Park look like a fish tank. His adventures with sharks and whales were much more dramatic than my encounters with rays and barracuda, and Cousteau had the coolest toys. When I decided to care about the environment—a prerequisite for a kid growing up in the predominantly white suburbs of the 1980s—I chose the ocean. My cause du jour while taking courses at Florida International University was trying to keep the oil industry from drilling in the Gulf of Mexico.

My environmentalism was sincere, and my concern for the ocean has only increased over the years. As the reports of the dire state of our ocean came out in the late 1990s, I became more interested in finding out how and why Americans come to care about places that are almost completely inaccessible, or at least physically inaccessible. I knew that my experience was not unique; many Americans experienced the ocean through television, movies, books, and magazines. And we relied on explorers like Cousteau to visit these regions and then interpret the landscape. The ocean that he imagined—that is, the ocean that he filmed, scripted, edited, and aired—was the ocean that we cared about. I wanted to know this process. Moreover, I wanted to know the history of ocean exploration. To what extent is the ocean that we know today a result of these popular ocean explorers?

These questions led me to combine a few historical subfields: the history of science and environmental history. The two fields have much to offer each other; my study here looks at a handful of popular oceanic naturalists throughout the twentieth century. None of them would really call themselves scientists; instead, they were natural historians caught in a world that decreasingly valued the work of naturalists. But they were explorers who shaped and developed the idea of the ocean in the American imagination. And they did so as part of a much wider history of American exploration. Indeed, when they saw the ocean, they often thought that they envisioned a wilderness frontier. When they created the pictures, stories, books, dioramas, and movies, they created a new ocean frontier. This process, I hope to show, was not without consequences. As we enter a world of accelerated ocean breakdown, we need to ask ourselves two questions: how did it happen? And what do we do now? This study contributes an answer to the first question by suggesting that turning the ocean into a wilderness frontier had a damaging effect. I can only guess at an answer to the second question, but breaking the cycle of creating new "last" frontiers would probably be a good place to start.

This project has been long in making, and I have piled up an impressive list of debts, only some of which I can recognize here. This book, first, is a proud product of public education. My thanks to the Faculty Scholars program at Florida International University and especially Darden Asbury Pyron, who first taught me the virtues of postmodernism. The University of Florida's history department offered valuable support while I was putting together my thoughts on Rachel Carson. It was there, under the tutelage of Eldon Turner and Betty Smocovitis, that I first learned how to think as a cultural historian. Thanks also to classmates Andrew Frank and Fritz Davis—more impressive colleagues I couldn't hope for. The University of Oklahoma's history of science program took a risk on me, and I hope that I haven't been a disappointment. My thanks to the entire faculty, especially Gregg Mitman, who taught me the importance of having a splendid advisor and friend during a student's doctoral years. I simply couldn't have done any of this without his encouragement. Thanks also to Donald Pisani and the rest of my dissertation committee, who gave me my first advice toward revising this book. Finally, my colleagues in the history department and the wider community at SUNY Plattsburgh have been supportive beyond words.

Early in the development of this project, I enjoyed the opportunity to spend two years thinking about the history of natural history in the twentieth century. A training grant from the National Science Foundation allowed five students and a "den mother" to travel around the country for a peculiarly delightful experience. My thanks to my fellow travelers for their patience and endurance. Special thanks go to the support of the grant's principal investigators: Gregg Mitman, John Beatty, and Jim Collins. And to our den mother, my close friend and colleague, Kevin Dann, thank you for teaching me how to live life properly.

Financial assistance was provided by the Milbauer Fellowship at the University of Oklahoma, a summer reading fellowship from the American Philosophical Society, and a dissertation research grant from the National Science Foundation. Smaller funding came from smaller grants and aid at FIU, UF, OU, and SUNY Plattsburgh. Thanks also to Graham Burnett for hosting the Science Across the Seas Princeton writers' workshop in the history of science. And special thanks to the marine environmental history group of the Sea Education Association, and the entire crew of the SSV *Corwith Cramer.* Of that group, Helen Rozwadowski, Michael Reidy, and Michael Robinson have been endlessly supportive as readers and deckhands. Also, Melissa Wiedenfeld and an anonymous reader gave recommendations on revisions that were especially valuable.

My friends and family have proven more important than I usually let on. Kim Nickerson's comradeship got me through graduate school. Elizabeth Hayes sat down and read my entire dissertation out loud—a hard one to repay. Melody Herr also lent a spirit of excitement to our histories of exploration. My parents, brothers, and sisters-in-law have had to endure my vacillations between excitement and outright despair. Thanks for your endless patience. And to my colleague and partner, Tracie Church Guzzio, your multiple readings of this manuscript saved the project from oblivion. But that is a comparatively small thing. My greater gratitude goes to you, Mirren and Lynn, for teaching me how to belong.

America’s Ocean Wilderness

Introduction

> Roll on, thou deep and dark blue ocean, roll!
> Ten thousand fleets sweep over thee in vain;
> Man marks the earth with ruin,—his control
> Stops with the shore.
> —Lord Byron, 1812

In 2002, listeners of National Public Radio heard Sylvia Earle, famed ocean explorer and then director of National Geographic's Sustainable Seas project, characterizing the practice of oceanic exploration. "So little of the ocean has been seen," she noted; "it is like the early days of exploring the American West."[1] At the same time as Earle was cruising through Caribbean waters in a one-person submersible, the directors of the Center for Marine Conservation announced that the organization was undergoing a concept change and would hence be known as the Ocean Conservancy. They were drawing from the analogous protectorate of landed spaces, the Nature Conservancy. The concept and name change, according to its directors, reflected "our new emphasis on conserving significant parts of our ocean as marine protected areas and as *ocean wilderness.*"[2] Similarly, Leon Panetta, chair of the Pew Oceans Commission, elaborated on this theme in a 2002 Earth Day address that took a page from Aldo Leopold, whose mid-twentieth-century call for a new "land ethic" helped raise awareness for the need for a more humane and ethical treatment of the land. "After all," Panetta exclaimed, "whether we live along the coast or in the heartland, the stewardship of our lands—and oceans—is our common national bond. This Earth Day, let us look beyond our parks, past the forests, and out into the sea with admiration and a new ocean ethic."[3] The goal of these statements was to extend the predominantly terrestrial nature of America's "wilderness ethic" to the oceans.

This mental process—the territorialization of the ocean—seems innocent enough. It comes as little surprise that humans make reference to the familiar to help understand the unknown, and that Americans specifically make reference to the western frontier wilderness to understand other frontiers like the ocean or outer space. American ocean explorers have been doing just this for at least the last century, and they are the subject of this book. Contemporary readers are probably most familiar with Jacques Cousteau, but other explorers figured prominently as well. From the start of the twentieth century, Roy Chapman Andrews, Robert Cushman Murphy, William Beebe, Rachel Carson, Eugenie Clark, and Thor Heyerdahl have also imagined the look, feel, sound, smell, uses, and abuses of an American ocean and reported back to us with slide shows, public talks, museum dioramas, world's fair exhibits, articles in newspapers and magazines, books, radio shows, movies, and television programs. Even today, we understand an ocean that has been shaped by explorers who have used the landed western frontier wilderness as their point of reference. Although it is important to know a little about this history, it is vital to recognize that the territorialization of the ocean may not be as innocent as it at first appears. But before we get to the moral of the story, we should consider the story itself—a narrative that begins at the start of the twentieth century, a time when many Americans were developing a new relationship with the ocean. Some Americans were beginning to view the oceans as a wilderness frontier that could replace the American West.

Most nineteenth-century Americans thought of the ocean as a *mare incognita*—an unknown space that flanked both sides of the continent. For instance, Washington Irving noted in *The Voyage* that "the vast space of waters, that separates the hemispheres is like a blank page of existence." The great sage of American sea writing, Herman Melville, described a seaman's thoughts on experiencing an ocean calm: "To his alarmed fancy, parallels and meridians become emphatically what they are merely designated as being: imaginary lines drawn round the earth's surface." Other times the ocean was a barrier, a wilderness to be feared and traveled as quickly as possible. Ralph Waldo Emerson, for instance, thought it "strange that the first man who came to sea did not turn round & go straight back again."[4] Nineteenth-century sea stories are about ships, storms, and whales; in them, the ocean became the setting for a human drama, and it usually represented the capricious force of nature or the will of God. Many American explorers took to the oceans during the nineteenth century, but for the most part, the thrust of American exploration was aimed elsewhere.

The nineteenth-century frontier was not oceanic; it was largely a century of terrestrial expansion, and that is where we find the explorers.[5] They mapped, cataloged, and prepared a continent for settlement. Geographers, botanists, zoologists, and federal scientists spent their time exploring rivers, breaking trails, and following railroad surveys. Lewis and Clark, Zebulon Montgomery Pike, Charles Frémont, Clarence King, John Wesley Powell, and John Muir imagined

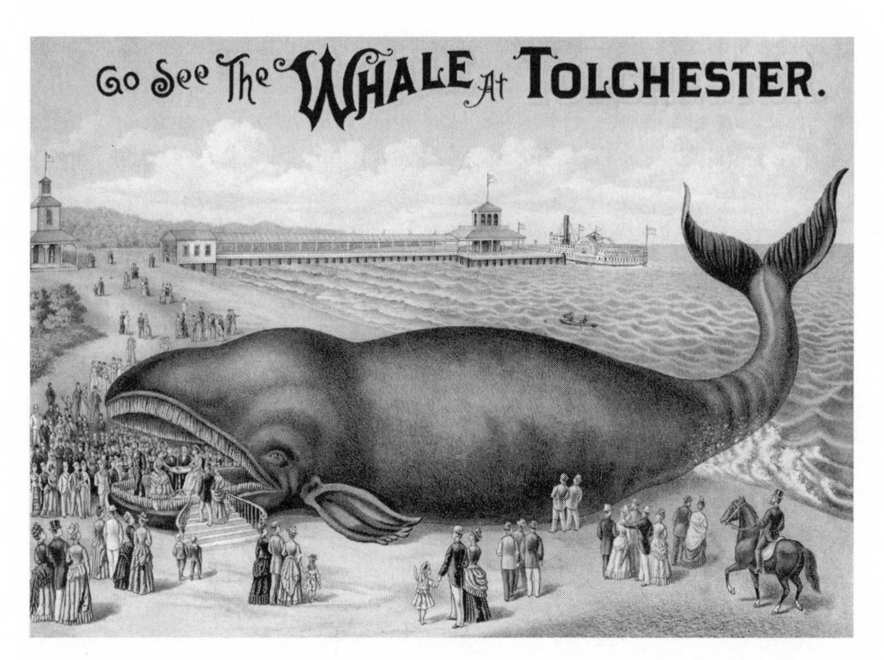

"Go See the Whale at Tolchester." Late-nineteenth-century baleen whale caught off Boston, transported to Tolchester, Maryland, and improved with parlor furniture for guests at the Tolchester Beach Resort. Courtesy of the Maryland Historical Association.

huge swathes of land and annexed them into an imperial archive so that the state and its citizens could make the landscape legible. The migrating population grew in size, and by the end of the century, the American consciousness entered a new phase. Frederick Jackson Turner's 1893 analysis of the closing of the frontier was a mere flash point of a wider anxiety that a postfrontier America was doomed to social and economic turmoil. Such sentiments spawned two important changes. First, a new ethic of scientific conservation sought to use the expertise of technocratic elites to manage the dwindling natural resources of a still young nation. The primary geography for this stratagem, according to Theodore Roosevelt, was the American West.[6] Second, American explorers began to search for new geographies beyond the continental margins. Explorers, previously hemmed in to the American continent, were unleashed at the end of the nineteenth century, and reports filtered in from Alaska, Hawaii, Guam, Puerto Rico, Panama, and the Philippine Islands.[7] A "splendid little war" against Spain transformed the ocean's status from a barrier to a conduit of empire.

About this time, in the icy waters of the North Pacific, Roy Chapman Andrews cut his teeth as a naturalist by investigating the biology of Pacific whales.

He operated out of the American Museum of Natural History in New York City, and his exploration brought him into intimate contact with the business of American and Japanese shore-station whaling, a brutally efficient industry that led to the decline of whale populations in the world's oceans. Andrews's work on Pacific whales between 1908 and 1912 demonstrates the movement of the American hunter into the ocean, but this was a "refined" class of hunters. Andrews was a member of a growing number of wealthy men who sought refuge from the decadent city in America's frontier wilderness. They were members of an elite eastern establishment whose wealth and status depended on the rule of efficient bureaucracy. Many of these hunters shared a notion of the frontier as a region full of commodities and resources valuable both for American industry and recreation. They also were becoming aware of the shortage of game species in what appeared to them an ever-shrinking frontier, a sentiment that spawned a desire for nationalizing resources through conservation laws. Finally, many of these hunters were either naturalists or included naturalists within their network of social exchanges. Theodore Roosevelt and members of the Boone and Crocket Club were representative of this hunter class. Andrews used this hunting ideology as he explored the natural history of whales in the North Pacific Ocean. In his writings, the ocean was filled with resources for human consumption—commodities in need of conservation through the expertise of a thoroughly modernized and efficient natural history. The ocean, for Andrews, was also a place to conquer and subdue with a hunter's cunning and skill. Andrews's stories, public presentations, and museum exhibits all conveyed these meanings, which were shared by a rising class of wealthy Americans who went to the ocean for a new frontier of manageable resources and sport.

Although Andrews conceived of the ocean as a place for modern business and sport, his colleague at the American Museum, Robert Cushman Murphy, painted a portrait of the ocean that was, by comparison, a bit more nostalgic. Murphy was the quintessential naturalist, an ocean explorer interested in describing the rich history of ocean life and the deleterious effects of humans subdued by the "myth of inexhaustibity." He was the nation's expert on oceanic birds, those aerial organisms that spend their lives drawing sustenance from the ocean. In Murphy's travels to the sub-Antarctic island of South Georgia, the guano islands of Peru, and the distant island nation of New Zealand, we find a naturalist who pays careful attention to the human and natural history of island territories whose fates are linked to the ocean. Through the use of biogeographical tools forged in the American West, he described a heterogeneous ocean in which life was linked to the peculiarities of climate, temperature, and ocean chemistry. Murphy saw the ocean as a true agent in the fate of nations, similar to the work of his favorite conservationist, George Perkins Marsh. He cautioned American audiences through books, articles, and museum displays that rampant overuse of the ocean would lead to certain disaster. Murphy based these warnings on the centuries of recklessness that the landed frontier had experienced under the witness of the pioneer's axe, which lent special relevancy to his argument.

Perhaps the most memorable ocean explorer of the early twentieth century was William Beebe, the eminent naturalist who climbed into a steel ball eight feet in diameter to be submerged to oceanic depths of up to a half mile. Hailing from an age that was witness to various firsts, such as the race to the Poles, the first transatlantic flight, and ballooning into the stratosphere, Beebe saw the ocean as a geography that provided similar physical challenges. But if Beebe was notable for his adventurous conquest of ocean depths, he was equally known for his portrayal of the dark and mysterious region of the earth that is the ocean deep. In the nineteenth century, romantic writers routinely brought the ocean into their repertoire of sublime landscapes, but they did so in a two-dimensional fashion along the plane of the ocean. Beebe's deep-water descents provided the venue for turning the ocean depths into another sublime region. But Beebe's articulation of an ocean sublime, in a manner similar to Clarence King and John Muir in the American West, was a function of his own identity as a naturalist making his way in a world largely dominated by modern laboratory science. A sensational character who was always in the public eye, Beebe managed the presentation of his own exploits in order to make his sometimes fantastic activities appear more scientific.

One of the most important themes of Beebe's ocean literature was that the planet earth was primarily a water planet, and the ocean affected the lives of landed humans in unpredictable and ubiquitous ways. I call this an "oceancentric" point of view. Oceancentrism took its cue from biocentrism, the idea that humans play only a small role in the drama of life. Rachel Carson took the biocentric philosophy of landed nature writers and forcefully extended the philosophy seaward in the 1950s. The scientific field of oceanography had experienced massive patronage during and after World War II, and so Carson thought it an auspicious time to write about *The Sea Around Us.* In this best-selling work, she presented an all-powerful and unpredictable ocean. Her work was at least partly motivated by her belief that both the destruction of the landed world in terms of land use and toxic pollution, and the corresponding geometrical increase in world population, would lead to the increasing importance of the ocean world. She couldn't have been more correct, and her job with the U.S. Fish and Wildlife Service gave her ample opportunity to view the expanding world of government science that subsidized exploitation of the postwar ocean. Carson, however, was a peculiar explorer; she found the beauty of the ocean in humble places along the American coast, and she called on her audience to do likewise. Although the lens of science could certainly provide expert knowledge of the ocean environment, it was just as certain that the typical American citizen could develop a sense of wonder.

While Carson was hard at work on *The Sea Around Us,* Eugenie Clark was serving as a naturalist for the Pacific Science Board in the distant American trusteeship of Micronesia. In 1952 she published a book that described her travels to the far Pacific and the Red Sea. *Lady with a Spear* created a new kind of ocean for the American public. It was a kinder and gentler ocean; full of beauty, wonder, and domestic wholesomeness, it was an ocean that Americans could experience for themselves with increased access to recently developed diving equipment. Just as American frontiers became popular spots for outdoor excursions among America's elite in the nineteenth century, so too did Clark help to create an ocean that could be similarly enjoyed and explored by the lay public. Whether she was cataloging poisonous fish in Micronesia or modifying the behavior of nurse sharks at Cape Mote Lab in Sarasota, Florida, Clark explored the ocean not as a conquering marauder but rather as a compassionate and even matronly nurturer. The domestication of the ocean, in Clark's hands, had everything to do with her own position as a naturalist in a field dominated by men, a fact made doubly complicated by the cult of domesticity that saturated postwar American culture.

Two of the most important shapers of America's postwar conception of the ocean were not themselves American, but their impact on the popular imagination should not be underestimated. Thor Heyerdahl's 1947 experiment to replicate the hypothesized migration of Native Americans to Polynesia captured a postwar American audience whose conceptions of the Pacific were products of death and destruction in the form of battleships and atomic bombs. Heyerdahl's idyllic float on a balsam raft was a kind of antitechnological narrative that emphasized the natural healing properties of nature and the ocean. Middle-class Americans began experiencing the Pacific as paradise after the war as Hawaii became a vacationer's destination; at a lesser cost, they could visit the Pacific at any one of the numerous tiki restaurants and bars that populated the postwar Sunbelt. If Heyerdahl's experience with the ocean promised a simpler move back to nature, Jacques Cousteau's ocean moved in the opposite direction, one that embraced a technologically savvy culture to mediate humans and the ocean. More than providing a window to undersea life, Cousteau's books, articles, films, and television series highlight an ocean populated by scuba-equipped man-fish, underwater scooters, underwater flying saucers, and housing units. Cousteau created an ocean that was easily explored and imminently habitable through the genius of science and technology. As Cousteau the explorer turned into Cousteau the environmentalist in the 1970s, he continued to look to the scientific and technological advances that would ameliorate the ocean's environmental problems. Taken together, Heyerdahl and Cousteau represent the oceanic extremes of a familiar terrestrial schism between people who fear science and technology as an environmental problem, and others who view it as the savior and redeemer of human activities.

Frontiers, the wilderness, and the West are all highly ambiguous concepts to use in systematic histories of space. At various times, and from different viewpoints, they could be conceived as regions of extraordinary danger to avoid; places of peril for seekers of adventure; lands to improve and settle; troves of natural resources and mineral riches; wildernesses to tame or conquer; landscapes in need of conservation for economic growth; a paradise necessary to exploit for human recreation; and ecosystems to preserve for the health of the natural environment. The definition of the frontier can include any one or a combination of these objectives, depending on a person's social and cultural identity. In the final analysis, it can only be said that the frontier is a mental conception of space that may bear little resemblance to the physical landscape. But that mental concept is important. It shapes policies such as military maneuvers, settlement patterns, environmental reforms, exploitative industries, social oppression, and racial inequity. How we think about frontiers can define our repertoire of behaviors toward these places and toward the organisms that inhabit them. If we wish to understand the contemporary issues that will clearly determine the fate of the ocean, then we must also understand how Americans have come to think about the ocean.

Today, we routinely hear reports of ocean pollution and depleted fisheries, exposés on dying coral reefs and endangered whales, and triumphant tales about preservation in the National Marine Sanctuaries and the Hawaii Ocean Preserve. These stories should have a familiar ring; they echo a century of efforts to conserve and preserve the American landscape. The plight of today's ocean, however, has been long in the making. When the ocean became an important geography in the American mind at the beginning of the twentieth century, so too began a process of utilization, degradation, and pollution, the consequences of which we are only beginning to realize. The central thesis of this book is that the ocean in the twentieth-century American imagination took on many of the characteristics that were typically associated with frontier territories: a trove of inexhaustible resources, an area to be conserved for industrial capitalism, a fragile ecosystem requiring stewardship and protection from "civilizing" forces, a geography for sport, a space for recreation, and a seascape of inspiration. These frontier meanings enjoyed wide circulation in the social imagination of the terrestrial frontier since the beginning of the nineteenth century.[8] Ocean explorers, like many Americans, enacted similar attitudes in their interactions with the marine frontier of the twentieth century. With a rapacity that would have stunned Lord Byron, one frontier replaced another, and the fate of both seem equally assured.

The Oceanic Hunting Grounds of Roy Chapman Andrews

Whales as Resource and Game in Postfrontier America

In the autumn of 1909, Roy Chapman Andrews, a young naturalist of the American Museum of Natural History, was far afield in Aikawa, Japan, on a solo exploration into the natural history of North Pacific whales. The Oriental Whaling Company had allowed Andrews to ride along on the fleet's steamers to observe the habits of various species of Cetacea. While working with the crews, Andrews developed a close bond with many of the Norwegian harpoon gunners whom he considered, no doubt with a twinge of envy, the archetype of the modern and courageous hunter. He later admitted that he "wanted to shoot a whale myself, but the gunners weren't very keen about it and I wasn't surprised. It wasn't sport to them. . . . But I was determined to add the biggest of all animals to my game list."[1] Andrews's persistence paid off, and he eventually convinced one gunner to let him try his hand. The ocean was as flat as a millpond when Andrews, clad in field khakis and campaign hat, manned a cannon that contained a hundred-pound explosive-tipped harpoon that was backed by three hundred drams of gunpowder. Despite carefully studying the hunting techniques of his Norwegian fellows, the first harpoon grazed off one unlucky specimen's head. After a few minutes passed, during which several of the crew scurried to reload the cannon, Andrews sent the second harpoon to the mammal's lungs. This particular whale was flensed and converted into oil and fertilizer before Andrews could pose with his trophy, but he would have other opportunities to stand next to whales while a camera captured the image of hunter and game.

This is a peculiar place to begin a story about a naturalist who is largely remembered for his leadership of the Central Asiatic Expedition of the 1920s, a large-scale expedition that failed in its goal to uncover evidence of human origins in central Asia, yet succeeded in discovering a large cache of dinosaur fossils in the dry sands of the Gobi Desert.[2] But Andrews cut his teeth as a museum naturalist with his cetacean studies in the cold waters of the North Pacific. The story may also seem to be a strange place to begin this analysis of Andrews as an American scientist at sea doing the serious work of exploring the natural history of whales. But the transformation of Andrews's scientific subject into an object of sport reveals the nature of his scientific practice and helps to explain the meanings behind his popular representations of both whales and the ocean. Andrews was a naturalist in one of the premier museums on the American continent, but he was also a hunter living in turn-of-the-century America. The ocean was his new frontier, the whale his game.[3]

Andrews was one of a breed of hunters new to early twentieth-century American culture whose cultural predecessors came from Europe's imperial hunting class. He was not a commercial hunter, nor was he a poor backwoodsy hunter out to secure food for the winter. He was adopted into a group of elite urban professionals, sometimes characterized as "the eastern establishment." The eastern establishment was a network of elite people and institutions that arose in the late nineteenth century in order to meet the challenges of America's new industrial order.[4] Hunters of the eastern establishment shared a common vision of the frontier wilderness, albeit a complicated and sometimes contradictory one. They viewed the frontier as both a resource of raw material for incorporation into the American economy and as a market for selling eastern manufactured goods. As much as they wished to incorporate the frontier into the eastern economy, they also defined the frontier, specifically the West, as the antithesis of—and antidote to—an industrialized civilization. A hunting expedition in the frontier promised certain rejuvenation, a sporting venture in which the sedentary male could test his mettle against a savage nature. Hunting stories and hunting trophies circulated, after the expeditions, in metropolitan men's clubs and were incorporated into a growing industry of popular culture that mythologized the West as an imaginary geography of frontier promise and adventure. These hunters were sometimes naturalists, or they brought naturalists into their circuit of social interactions—an easy task given that America's scientific elite occupied the same metropolitan spaces as America's financial elite. The eastern establishment was equally captivated by the practice of exploration. Many of them toured the U.S. West as a kind of rite of passage; others became the source of funding for grand explorations and expeditions into unknown territories.[5]

Part of the animus that energized the hunter's ideology was a spirit of "frontier anxiety," a widely shared concern that the "closing of the frontier" presented serious challenges to American culture and economics.[6] Hunters of the eastern establishment reacted in a number of ways. They sought to preserve an ostensibly pristine and undeveloped fragment of the frontier West through the establishment of parks and preserves. They also attempted to use the tools of science and technology to efficiently conserve the nation's remaining natural resources.[7] The goal of efficiency became something of a business credo in turn-of-the-century America. In this capacity, the scientific bureaucrat became an instrumental tool for managing both industry at the core and resources at the periphery.[8] Finally, the eastern establishment hunter, anxious over the closing of the frontier and all that it portended, became an instrument of the spirit of American imperialism in his search for new extracontinental frontiers to conquer. The icons of this hunter ideology were, of course, Theodore Roosevelt, the members of the Boone and Crockett Club, or—as was the case for Andrews—members of the New York–based Explorers Club.

Naturalists of Andrews's social standing entered the eastern establishment through a side door. Andrews had a humble upbringing in the Wisconsin backcountry. He was an average student, an adequate observer of nature, and had a fondness for hunting. After graduating from Beloit University, he headed east to apply his hunting skills to a higher purpose. Charismatic people like Andrews were often invited in to America's financial and social elite as a form of entertainment for the industrial and banking social classes. The American Museum of Natural History was a key locus of this culture in New York City. The building itself was erected as a monument to the extraordinary wealth of an industrializing America, and its presidents, directors, and staff continuously entertained a network of wealthy patrons who often funded museum science.

New and wildly profitable industries that embraced the principles of scientific management generated the capital that built the architectural symbols of wealth, grandeur, and empire like the American Museum of Natural History, the apotheosis of all that was civilized and modern. In characterizing the rich social texture of museums like the AMNH, one historian has noted that "local boosters, proponents of popular education, conservationists, imperialists, and lovers of nature found common ground in promoting the construction of [these] 'cathedrals of science.'"[9] The AMNH, an institution that often turned exhibited organisms into displays of power, wealth, and values, was a public space that embodied urbane ideals, especially those of conservation and hunting.[10] This reading is evident even today at the AMNH's Roosevelt Memorial. As if linked to the vast western expanses that enabled the accumulation of Gilded Age wealth, the museum had since its inception in 1869 prioritized the collection and display of terrestrial organisms. Then, in 1907, the museum moved to incorporate the ocean into their nature archive.[11]

At the urging of the museum's president, Morris K. Jessup, George S. Bowdoin—a partner of J. P. Morgan and trustee of the museum—donated $10,000 for the construction of a life-size model of a blue whale. The task fell to an ambitious Roy Chapman Andrews, who had been recently hired at the museum to perform odd jobs, and the new staff taxidermist, James Clark. The two spent some eight months constructing a 76-foot model made of iron ribs, wooden frame, and papier-mâché. Some Long Island fishermen successfully harvested a North Atlantic right whale in Amagansett Bay, and so the museum workers paused to spend three days battling freezing winds to secure the entire skeleton and the baleen, purchased with $3,200 of Bowdoin's donation. The brief expedition gave Andrews practical experience with whale anatomy that was helpful in constructing the model. When finished, the blue whale facsimile represented "a notable forward movement in the policy of the museum." As opposed to stuffing a carcass with arsenic paste, this policy used new taxidermy techniques to create educational exhibits that were a "correct representation of the animal as it lived."[12] Despite such accolades, the model was suspended over the Hall of the Biology of Mammals gallery from wires that might recall the mechanical apparatus then used to move the whaling industry's organisms over a shore-station slip. It was positioned horizontally, not as if floating lazily, but rather assuming the position of a whale being drawn out of the ocean for butchering. In short, it conveyed the message that the ocean was filled with resources ripe for the harvest.[13] With two whale projects completed—the blue whale model and the retrieval of the North Atlantic right whale—and with $1,000 of Bowdoin's money left over, Andrews requested permission to further his career by undertaking fieldwork in the North Pacific Ocean.

EASTERN PACIFIC WHALING AND THE CONSERVATION OF WHALES

Andrews's first substantial experience as both an explorer and cetologist began with his April 24, 1908, departure for the Pacific whaling stations on the coast of Vancouver Island. He went to secure photographs, notes, and measurements relating to the largely unexplored cetaceans of the Pacific. "The entire scientific knowledge of these forms," Andrews noted, "rests on the observations of Captain Scammon, made more than thirty years ago, which have never been verified." He also wanted to add several skeletons to the museum's collection of Cetacea.[14]

The success of this expedition and those that followed relied on the generosity and goodwill of Pacific whaling companies that had only recently

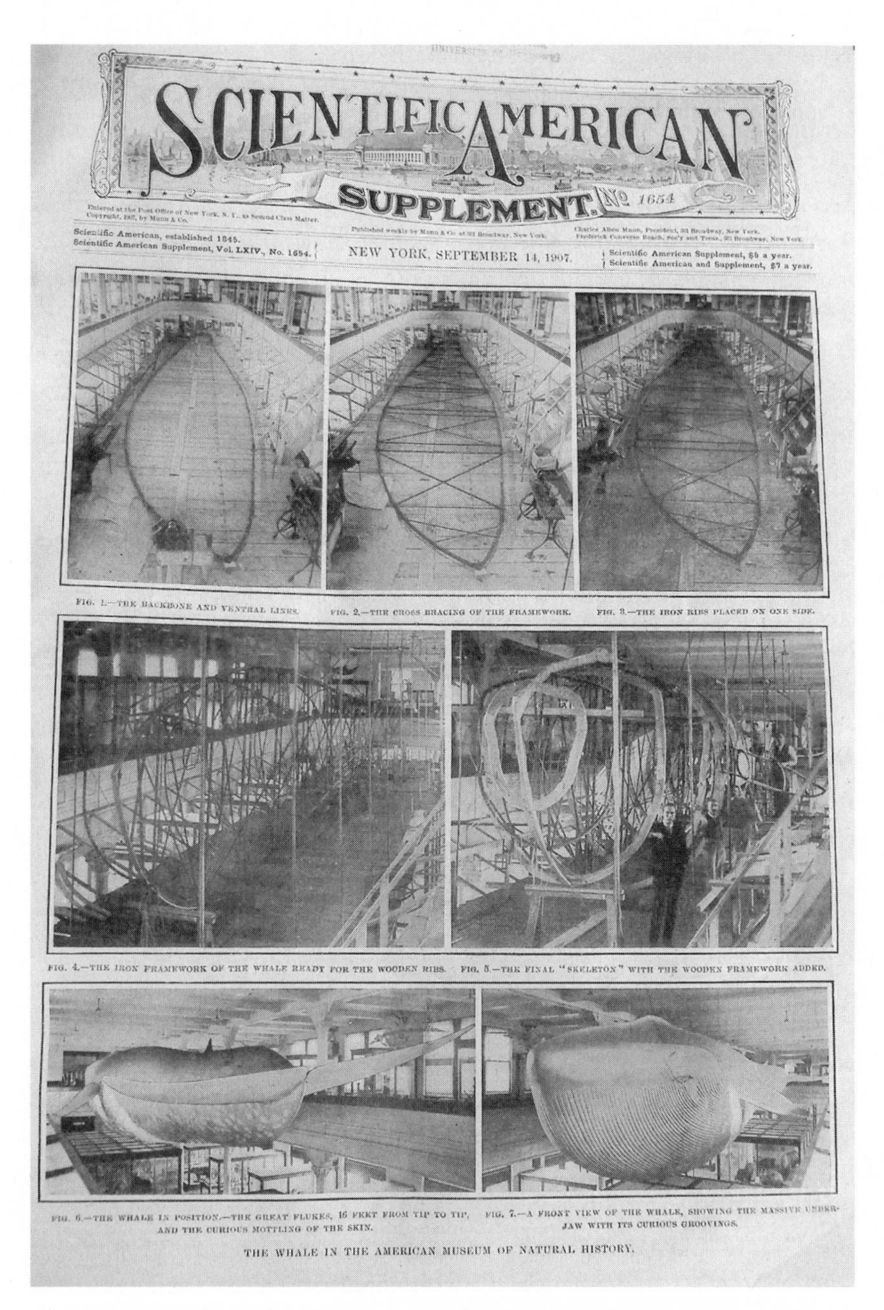

SCIENTIFIC AMERICAN SUPPLEMENT. № 1654

Entered at the Post Office of New York, N. Y., as Second Class Matter.
Copyright, 1907, by Munn & Co.

Published weekly by Munn & Co. at 361 Broadway, New York.

Charles Allen Munn, President, 361 Broadway, New York.
Frederick Converse Beach, Sec'y and Treas., 361 Broadway, New York.

Scientific American, established 1845.
Scientific American Supplement, Vol. LXIV., No. 1654.

NEW YORK, SEPTEMBER 14, 1907.

Scientific American Supplement, $5 a year.
Scientific American and Supplement, $7 a year.

FIG. 1.—THE BACKBONE AND VENTRAL LINES.

FIG. 2.—THE CROSS BRACING OF THE FRAMEWORK.

FIG. 3.—THE IRON RIBS PLACED ON ONE SIDE.

FIG. 4.—THE IRON FRAMEWORK OF THE WHALE READY FOR THE WOODEN RIBS.

FIG. 5.—THE FINAL "SKELETON" WITH THE WOODEN FRAMEWORK ADDED.

FIG. 6.—THE WHALE IN POSITION.—THE GREAT FLUKES, 16 FEET FROM TIP TO TIP, AND THE CURIOUS MOTTLING OF THE SKIN.

FIG. 7.—A FRONT VIEW OF THE WHALE, SHOWING THE MASSIVE UNDER-JAW WITH ITS CURIOUS GROOVINGS.

THE WHALE IN THE AMERICAN MUSEUM OF NATURAL HISTORY.

The 1907 cover of *Scientific American* here describes the phases of construction of Roy Chapman Andrews's model of a sulfur-bottom whale that hung in the American Museum of Natural History until 1969. The accompanying article notes that when captured, the "whale is towed ashore, where in several places machinery has been installed for cutting up the carcass. Little goes to waste." From *Scientific American Supplement* 1654 (September 14, 1907), cover and 162.

developed operations in the North Pacific. Before his departure, Andrews contacted Captain Balcom of the Pacific Whaling Company that operated out of Secart, Vancouver, and requested permission to observe whales and whaling operations from both steamers and shore stations.[15] He assured Balcom that his "investigations would not interfere in any way whatever with the work at the station."[16] Andrews's research agenda relied heavily on the practices of the modern whaling fishery; he thus carefully forged a conciliatory relationship with industrial interests.

The first stop on Andrews's itinerary was Vancouver Island, where the Pacific Whaling Company managed two shore stations near Barkley Sound and Kyuquot Sound. Beginning in 1905, these stations hunted humpbacks, blue whales, and occasional finbacks, which were converted into oil and fertilizer. The Pacific Whaling Company represented a reallocation of fishing capital at the start of the twentieth century. Previously, America's whaling fleet had been based in New England, though the Pacific was no stranger to such whalers. The new Pacific Whaling Company, as its name implies, operated out of the Pacific Northwest and the processed oil and fertilizer supplied the market in the western states. The company had expended no small amount of capital in erecting engine houses, wharves, bunkhouses, offices, and the machinery required for refining whales.

The Barkley station was situated in a small bay not far from the open sea where the whales were hunted, killed, and then towed back to the station for processing. Every morning, the station's two steamers went to sea. With crews of twelve, these vessels were about a one-hundred-ton burden, ninety to one hundred feet in length, and round bottomed to facilitate speedy manipulation. Stations like the Barkley site, which were quickly constructed and often just as quickly abandoned, resembled the stereotypical western ghost town that flowered in times of prospecting, only to disappear when the wealth of a region had been exploited.[17]

The Barkley station was representative of a fairly new and rapidly growing industrial process designed to efficiently harvest large numbers of whales through the use of modern technology. Around 1870, two technological innovations breathed new life into the stagnant industry. The first was the wide application of steam power, which harnessed the sun's latent energy stored in fossil fuels and enabled whalers to chase down the faster-swimming species—rorquals, bowheads, and finbacks—instead of chasing after sperm whales while under sail. The second innovation was of Norwegian design. In 1868, Svend Foyn constructed an explosive-tipped harpoon gun that would allow whalers to do the large part of their work aboard the main ship. This invention cut down the amount of time required for securing a whale; whaling became a matter of getting close (which was no easy task for fast-moving rorquals) and firing a harpoon from a cannon. The whale was often killed instantly when the bomb exploded in the animal's lungs.

A good day brought in as many as twelve or fifteen whales, which were towed to a shore station where a team of men took their positions on a line. As the whale was towed up the landing slip, the carcass was quickly dissected and converted into marketable products. The entire process, like the industrial slaughterhouses of Chicago, was remarkable for its speed and efficiency, and just as Chicago was responsible for transforming the environment of its Midwestern hinterland, so too did shore stations alter the biological community of their oceanic hunting grounds.[18] Shore stations also made efficient use of parts of the whale that were formerly neglected. In the old style of whaling, workers collected blubber from the whales (usually sperm whales) and then set the remainder of the carcass adrift. In contrast, shore stations of the late nineteenth century were much more effective in utilizing the entire carcass. Norwegian and British whalers, starting in the 1870s, used the newer techniques with terrifying celerity. America would boast similar outfits in New England harbors in the 1890s. As a result, whalers quickly decimated cetacean populations, and in the first decade of the twentieth century, the entire Atlantic industry moved to the South Atlantic, where the whales off South Georgia produced tremendous, though temporary, wealth. As the Pacific Whaling Company illustrates, American whalers also turned their attention to Pacific species, establishing similar operations in Alaska and British Columbia.[19]

At the end of May, Andrews made his first observations of whaling operations on the steamship *Orion.* After storing his equipment below deck, he began engaging the captain, gunners, and crew in conversation. Perched on a great coil of towing line, he spent much of his time asking questions about the specific species of whales they encountered, and he prodded the crew for interesting stories of close encounters and hairbreadth escapes. Immediately after a cry from the crow's nest indicating three humpbacks off the port bow, Andrews scurried to gather his camera, notebook, and pencil. He stood next to the captain, who taught him how to identify whales by observing the shape of their spouts. Still without his sea legs, Andrews had great difficulty taking notes, fiddling with his camera, and trying to keep dry at the same time. Every time the *Orion* closed in for a shot, the whales promptly dived. The captain predicted where they would rise again and adjusted the steamer's heading. This game of cat and mouse continued as the whales' dives became shorter and shorter. The *Orion* chased after this group of three humpbacks for two hours before striking one with the cannon. On this excursion, and the many that would follow, Andrews took notes on the behavior of whales: spouting, diving, swimming, feeding, and mating. However, the knowledge he gained from the crew should not be underestimated; indeed, many of Andrews's popular and scientific articles use anecdotal evidence supplied by gunners and captains.[20]

After several weeks of work on Vancouver Island, Andrews went to a station on Admiralty Island, Alaska, operated by the Tyee Company. He spent three weeks off the Alaskan coast; his work at this site yielded far superior photographs, mostly as a result of calmer seas. While conducting his research from the deck of several whale boats, Andrews amassed "a magnificent collection of photographs taken outside without having to go through the terrors of seasickness. . . . I got one of the harpoon in the air, and another after it had hit the whale; and photographed Humpbacks and Finbacks from every conceivable point."[21] As clearly stated in the title of his popular account of whaling, *Whale Hunting with Gun and Camera* (1916), Andrews's camera functioned as a substitute for the harpoon cannon. As a steamer neared a humpback on his first Alaskan sortie, he quickly removed himself and his camera to the harpoon deck. After some anxious minutes, "the roar of the gun almost deafened me and instinctively I pressed the button of the camera."[22] Andrews's instinctive impulse thus linked photography and hunting as analogous practices in the work of natural history. "Whale hunting with a camera," he wrote in a beautifully illustrated article for *World's Work* shortly after his return, "is a royal sport."[23]

When Andrews was not taking notes and photographs aboard whaling ships, he spent his time onshore measuring, photographing, and describing the daily catch. As humpbacks, sperm whales, and blue whales moved up the slip, Andrews would busy himself measuring their features and inscribing the details into tables. He noted the variation in color, dissected their stomachs, and even found time to scrutinize their skeletons. During his two months in the Pacific, he observed over one hundred whales; before the age of the shore station, he often remarked, a naturalist might have been content to observe only a handful in a lifetime.[24]

After his return to New York, Andrews continued to call on whaling personnel for information. Shortly after his Vancouver expedition, he wrote to an official of the Pacific Whaling Company, "I am wondering how the Page's Lagoon station is turning out this winter, and whether you found that the whales which were there really had been killed off, or whether new ones were appearing."[25] Andrews was becoming aware that modern shore stations were severely, if not irreparably, affecting the populations and even the behavior of Pacific cetaceans. The decreasing numbers of these animals were a double loss. Out of sheer pragmatism, Andrews thought that the dwindling populations

Roy Chapman Andrews's photograph of a harpoon cannon firing at a finback whale during his expedition to Korea, 1911–1912. After a successful shot, the whale was towed back to shore stations for dismemberment. Courtesy of the American Museum of Natural History.

inhibited the naturalist's potential for discovering the evolutionary histories of cetaceans. But it also signaled a loss to the whaling industry that was rapidly overharvesting its natural resources.

Andrews expressed his concern in an interview that was published in the *New York Times* upon his return. But it was actually the museum director, Herman Bumpus, who first noted that North Pacific whale populations were reaching the point of "commercial extinction." Bumpus was "convinced that the whale is rapidly becoming extinct, and in a comparatively few years will take its place beside the dinosaur and the three-toed horse of the shadowy prehistoric ages."[26] Later, referring to the stations he visited on the eastern Pacific coast, Andrews would echo Bumpus's concern by remarking that "after two years work, practically all of the whales were killed off and no others have come in these Island waters."[27]

Perhaps Andrews did not foresee the consequences of bringing this subject into popular discourse. Upon his return, reporters harried Andrews for reports of Alaska, whales, and whaling. Andrews noted that Pacific fisheries "conducted the whaling in such a way, that within a very few years all the whales would be killed out and extinct on this coast." Shareholders of the Pacific Whaling Company, incensed by these reports, instructed management to refuse all requests for specimens and scientific use of their facilities. They claimed that their commercial interests had been compromised and that the value of their stock had been damaged by Andrews's statement.[28]

Andrews responded to the accusation by placing blame on newspaper reporters. "I was besieged with reporters on my return and when I did not give them the sensational stuff which they desired, they made up the remainder from their own ideas. . . . You must know," he went on, "that I would not intentionally do anything to injure the interest of the men who made possible the work which I was able to do last summer."[29] Such problems called for delicate diplomacy between Andrews and the patrons who supported his work. Andrews's response to the Pacific Whaling Company also demonstrates the murky relationship between the sensationalism of adventure and the higher calling of science. He claimed that he was reluctant to give the newspaper reporters sensational copy; he wanted to hew close to an objective report of his scientific findings. The youthful Andrews was beginning to appreciate and cultivate the notion that popular attention to his natural history must include some element of sensationalism. The event set into motion an implicit strategy that would characterize Andrews's career as a popular explorer: the need to carefully balance adventure with science.

The lesson was an important one for any naturalist interested in undertaking far-flung expeditions with high price tags. Publicity and popularity were essential to an explorer's success in Progressive America.[30] Andrews's first popular treatment of Pacific whaling appeared in *World's Work* late in 1908. The editor, a vice president of the American Telephone and Telegraph Company, turned down the first draft and suggested that Andrews write less scientifically and instead focus on the adventure of the story itself. The final product was a well-illustrated article that told a simple story of how Andrews made a sport out of hunting whales with his Graflex. Andrews then turned the adventure into a public lecture that included over a hundred colored lantern slides and eight hundred feet of motion picture film. He first lectured at the Five Points Mission, where he "tried to make them feel and hear the rush of the sea, the roar of the gun and the thrill of the hunt."[31] Andrews's lectures that winter were transforming his audiences' conception of whales into game that lived within a geography of manly sport. Andrews pulled his audience in with a charisma that would soon win him audiences with New York's social and financial elite.

The expedition to the eastern Pacific publicly marked Andrews as an adventurous naturalist in the same tradition of explorers like Robert Peary, who at the same time was racing toward the North Pole. The delicate balance between adventurer and scientist was articulated in a *New York Times* article that appeared shortly after Andrews's return: "Primarily, [Andrews] is a scientist, but his love of adventure in unknown worlds of his science runs a close second."[32] Science and adventure, according to the author of this piece, were mutually exclusive terms that could be incorporated into the body of a single person, but at the same time remain distinct practices. The article emphasizes the manly characteristics requisite for the work of exploration. When we delve into Andrews's past, we find that "at the university he used to play baseball and take a hand at athletic sports generally. That gave him a lithe, muscular body—just what he needed to chase a bull whale to his habitat and look him squarely in the eye." That Andrews was on the chase is a theme that permeates the article. Theodore Roosevelt provided the archetypal narrative of the explorer's chase or hunt. Shortly before Andrews's expedition to the North Pacific Ocean, readers of the *New York Times* could find daily reports of Roosevelt's latest trophies hunted down in the heart of the African Congo. Just as Africa was fashioned into a new province for testing the manly strength of Roosevelt, so too did Andrews represent the ocean as something of a playground in a geography of virility. He sought to preserve the virtues of rugged masculinity so as to fend off the decadence of the sedentary businessman.[33] Moreover, whales functioned much like the lion, elephant, and okapi as a most noble and prized game. One worthy specimen, a sulfur-bottom, Andrews told the *Times* reporter, "reared and charged furiously and kicked up a perfect tempest in the sea when the whaler got into his front yard. He was madder than an exposed politician and didn't care a campaign whoop who knew it."[34] The taming of this beast, both by the whaler and the naturalist busy taking pictures, was a testament to the courage and daring of the scientific adventurer.

Also of note is the article's coverage of the whaling industry. Andrews learned that "there is no longer any romance in the whale-hunting industry. It has fallen hopelessly into a deadly dull business. Yes, the old Nantucket whalers have had their day. The industry is . . . devoid of the romance and poetry of other days." This is most likely the sentiment of the reporter; Andrews's attitude was less nostalgic. Take, for instance, his commentary on the epitome of nineteenth-century whaling, Herman Melville's *Moby-Dick.* He was

> surprised to see how much all New Bedford people seem to think of [Melville]. Some parts of it seem to me to be good, but as a whole I think it is uninteresting and tremendously inaccurate in many parts. The story would be good if about one-third of it were cut out and Melville had learned something about whales before he tried to write it. He seems to know nothing whatever about the anatomy and very little of the natural history of whales.

> . . . I presume its appeal to New Bedford people is because it does give a good picture of New Bedford during the time of whale days. From the standpoint of the whale, the "Cruise of the *Cachalot*" by Bullen is a good deal better, I think, and he knows a good deal more about whales than Melville did.[35]

Clearly, Andrews was not captivated by the history and lore of nineteenth-century American whaling. While others would characterize modern shore-whaling as a "deadly dull business," Andrews was much more sanguine about the efficiency-minded practices of modern whaling.

But as Andrews considered the efficient harvesting of these organisms, he began to express greater concern over their imminent demise. He was by no means a great conservationist. In his autobiography, he playfully minimized his conservationist efforts: "Always I have been a believer in conservation. I conserve old hats, old shoes, old pants and particularly old slippers, to the disgust, I may say, of my wife."[36] And it is true that during the apex of his career as an explorer of Central China, conservation concerns were far from his mind. But during this early period, he did make some efforts that fit into the context of business-oriented Progressive conservation. Andrews viewed the practice of shore whaling as brutally efficient in the short run but hardly sustainable over time. When he did talk about the possibility of extinction, he used the phrase *commercial extinction,* by which he meant a decrease in populations of whales at which point their pursuit would no longer be profitable.[37] Although it is unclear whether Andrews actually advised station managers on the best way to perpetuate the economic viability of the fishery, he demonstrated a concern that whaling was being conducted without the wisdom of scientific management.

He made some effort at bringing this issue to the attention of politicians.[38] Andrews drafted a bill, apparently never introduced to the floor of Congress, to prohibit shore whaling off the coast of Alaska without a license issued from the governor and approved by the commissioner of fisheries. Factories should be conducted "in such a manner that no injuries or deleterious matter will be introduced into any public waters." Factory equipment should be approved for their efficiency based on industry standards. In order to prevent excessive clumping of shore stations and the inevitable whale population declines that followed in their wake, each station was to be separated by a distance of two hundred fifty miles. Finally, in contrast to the common practice of each station using numerous steamers, Andrews thought that a "one factory-one steamer" restriction would prevent overharvesting. In short, Andrews was attempting to use his expertise as a naturalist to create policy that would bring the principles of scientific efficiency to the whaling industry. Although Andrews no doubt believed that the extinction of various whale species would be a tragic loss to the oceans' natural history, the wording of his bill demonstrates that, in his opinion, ocean dwellers were resources destined for consumption, and only scientific management could ensure the sustainability of those resources.[39]

Andrews was not the only naturalist concerned with the problems that shore stations presented to whale populations, though other naturalists chose to focus on international diplomacy over technological restrictions. In November 1909, the U.S. Fisheries Commission held a conference in which California naturalist David Starr Jordan and the commissioner of fisheries, George Bower, laid a recommendation on the desk of the U.S. Secretary of Commerce and Labor, Oscar Straus. They called for an international conference with a view to regulating the killing of seals, whales, and all other mammals of the sea.[40] International cooperation was hardly new to conservation in the first decade of the twentieth century. Beginning in the 1890s, international conservation between the United States and Canada became a concern to politicians, naturalists, and businessmen alike; inland fishes and the North Pacific fur seal were the dominant concerns at that time.[41] Despite this precedent, whales did not become subjects of international conservation until the late 1930s. Nevertheless, Commissioner Bower singled out the specific stations visited by Andrews in 1908 as having a particularly devastating effect on whale populations: "In recent years the establishment of whaling stations on the British Columbian Coast and one in Southwestern Alaska has caused a rapid decrease of whales off that coast. One or more of the species are so near extinction that a closed period really ought to be provided. This, of course, could be brought about by international agreement."[42]

The importance of a broader policy of marine conservation that included the ocean's fish populations also made some headway in the 1910s. European fishers were aware of the possibility of overfishing the ocean, which led to the creation of the International Council for the Exploration of the Seas in 1902.[43] American fishers were a little behind in this realization, and it was not until the advent of ocean trawl-net fishing that Congress began to at least consider taking steps to regulate marine conservation. "In search for raw material out of which valuable by-products come," remarked an editorialist, "certain corporations are . . . diminishing seriously the food supply of cities along the Atlantic coast." Linking the potential plight of the ocean with the fate of the land, the editorial noted that "as with the forests on land, the supposition has been that nature's supply is inexhaustible." There was thus a great need for the "conservation of marine food."[44] But dwellers of the ocean, whales included, had more than a utilitarian value to some.

No one was more fervent about the plight of whales than Andrews's colleague, Frederic A. Lucas, then a naturalist and director of the Brooklyn Museum of Natural Science. Lucas clearly linked the problems of whale overfishing with the changing practices in the industry. "The old sailing boats equipped with hand harpoons," he stated in an almost nostalgic editorial, "have been displaced by shore stations near the breeding grounds, from which, at a signal from the lookout, swift steam vessels dash, hurling their shot harpoons and bomb lances at the cetaceans." Such practices would clearly result in nothing short of extinction. Perhaps more alarming was the proliferation of shore stations throughout the world—the Pacific Coast, Patagonia, Iceland, the Faeroe Islands, New Zealand, and South Africa. "Something must be done," Lucas continued "by international agreement, or the fisheries will destroy themselves and the whales with them."[45]

Lucas expressed similar views in several articles and letters that appeared in the *Scientific Monthly Supplement.* Perhaps the most interesting characteristic of these pieces was the philosophical tone of his concern. It was more than just an industry falling on hard times; the overfishing of whales was another indication that nothing "can escape the all-grasping and all-powerful hand of man." Furthermore, man's grasp on nature had become increasingly tighter in the modern period. In contrast to the "old method" of whaling that required incredible amounts of capital, labor, and danger, "steam and the whale gun have changed all this and largely destroyed the romance of whaling. There is little excitement save in the chase, which is often prolonged." Shore stations became a symbol for the danger posed to ocean dwellers when capitalism joined hands with technological advance. Romance was replaced by efficiency, and in the latter's wake, nothing was more certain that the "passing of the whale."[46]

Lucas also framed his concern in terms of the myth of inexhaustibility: "Another fallacy was the belief that the supply of whales was practically limitless and that one might 'slay and slay and slay' continuously. There is not a more mischievous term than 'inexhaustible supply,' and certainly none more untrue." Lucas, using the robber baron's lingo, believed that "man is recklessly spending the capital nature has been centuries in accumulating and the time will come when his drafts will no longer be honored." The history of Newfoundland shore stations had made abundantly clear, according to Lucas, that the worldwide proliferation of such methods meant the possible elimination of an entire biological order. Lucas believed that the only solution to the problem was through strict regulation of the industry through licensing and international collaboration.[47]

Lucas heavily influenced Andrews's thoughts on the conservation of whales, but there were major distinctions between them. First and foremost, Lucas was troubled by the extinction of an order of fauna. Simply put, it would be better for the oceans if these leviathans continued to inhabit them. He made a natural link to the then recent success in saving the bison from oblivion. Now, he proclaimed, it is time to rescue the bison's "warm-blooded cousins from the ocean plains."[48] Of course, this was a value judgment that operated on aesthetic and emotional levels. It tied into an environmental sentiment that pervaded turn-of-the-century American culture in such manifestations as the Sierra Club, the American Ornithologists Union, and the Boone and Crockett Club. These organizations worked to preserve—for diverse reasons—some vestige of the natural landscape that would inevitably be threatened at the hand of American expansion. This sentimental motive is absent from Andrews's thoughts and writings. He falls into the realm of Progressive conservation in which nature must be scientifically managed for American industry. To be sure, these two traditions often converged. With the exception of the nod toward international cooperation, Lucas's call for regulations was identical to Andrews's. The difference, however, was its motivation: where Lucas was concerned with the future of whales, Andrews was concerned with the future of whaling.

ASIAN FISHERIES AND THE GAMING OF OCEAN FAUNA

Andrews intended to make a second Alaskan visit in the summer of 1909, but a long trip to the Saint Lawrence River on a quest for beluga whales for the New York Aquarium made a western trip impossible. Andrews's next exploration of Pacific Cetacea concentrated on the whale fisheries of Japan and Korea. What would be his longest and most successful whaling expedition began with an invitation from the U.S. Bureau of Fisheries to serve as a naturalist of porpoises on an expedition to the East Indies aboard the exploring ship *Albatross*. The Bureau of Fisheries' work in the Pacific dated back to the 1880s when the federal government began sponsoring hydrology, natural history, and cartographic surveys of Alaska, mostly to seek out new areas to relocate the failing salmon and halibut fisheries of the Pacific Northwest. Shortly after the Spanish-American War, Theodore Roosevelt and a number of naturalists called for similar surveys around the Philippines.[49] The *Albatross* was not commissioned to do so until 1909.

Early in September 1910 Andrews left Seattle for Yokohama aboard the S.S. *Aki Maru*. The Far East captivated him, especially its social life. Andrews made good use of the network of quasi-colonial social clubs that functioned as safe harbors for world travelers. In Yokohama, he was fond of lodging at Number Nine, "the most famous house of prostitution in the world." Similarly, his work in China, Korea, and the Philippines often included visits to the British, American, and French army and navy clubs, all institutions "important to the eddying currents of cosmopolitan life which flowed through its doors." These clubs were foreign analogs of the many genteel social clubs for the elite that peppered the metropolitan landscapes of New York and London. They served as clearinghouses for information and provided important headquarters for explorers of Andrews's kind.[50]

The *Aki Maru* left Yokohama and traveled south across the East China Sea to Shanghai, then through the Formosa Strait and on to Hong Kong. Here Andrews left the *Aki Maru* for the S.S. *Tamin* and traveled across the China Sea to Manila, where for two weeks he waited for the *Albatross*'s arrival. Once safely aboard, Andrews played no role in the many dredges that were the daily work of *Albatross* naturalists working for the federal government; every time the crew conducted a dredge close to an island, he would remove himself to the shore to collect birds and mammals. Andrews did not restrict his hunting to land animals, however. Aboard the *Albatross,* he made targets of dolphins and small whales with the use of a bomb gun that fired an explosive from a rifle.[51] But the quick stop-and-go pace of oceanographic surveying made it impossible for him to do any work in serious detail. After completing a survey off the coast of Celebes, the *Albatross* headed northward and dropped Andrews off in Nagasaki before heading back to America.

Andrews was supposed to be on board the *Albatross* and bound for America, but a fortunate walk through a Yokohama fish market changed the course of events. He noticed that merchants sold whale meat in great quantities, and he soon realized that Japanese whaling stations would offer opportunities to study whales similar to those in Vancouver and Alaska. This was a lucky turn of events; the next six months would be his most successful as a whale naturalist. The American consul arranged an interview with officials of the Oriental Whaling Company based in Shimonoseki. The president of the company was kindly disposed to Andrews's work and granted him permission to visit the shore stations.

He first stopped at a station at Shimidzu on the island of Skoku, but the catches were disappointingly meager, so he moved on to Oshima, where, as he reported to Joel Allen, he "could deliver the goods this time in unlimited quantities. . . . This is absolutely the time of all others to load up with whales. The Company has given me *carte blanche* to get anything I want and they treat me like a king. . . . Never again will we have a chance to get so much for so little money."[52] Allen enthusiastically replied, "such an opportunity for observation and field research in this line as you now have has probably never before fallen to the lot of any cetologist, and you deserve your good luck and I am sure will make the most of it."[53]

Andrews was increasingly spending more time on shore than aboard whaling steamers. He observed the operations of the shore stations and absorbed the niceties of Japanese culture. Like his work in eastern Pacific stations, he took descriptions and photographs of the daily catch. When larger specimens arrived, Andrews went about the work of cleaning and preparing the skeletons for shipment back to New York. At Oshima, he secured a seventy-nine-foot blue whale, a forty-six-foot sei whale, and a twenty-six-foot killer whale. A large sperm whale was also on his list of desired specimens, and this he sought from the company station at Aikawahama in the province of Rikuzen. He stayed more than three months accumulating data both ashore and aboard station steamers. A prize sixty-foot sperm whale was finally taken as a special favor to Andrews. Indeed, Andrews received many favors from company officials, who occasionally issued orders to help him fill out his list. In return for this service, company officials received miniature models of the whales they had given Andrews—models laboriously crafted by Jimmy Clark back at the American Museum.

The specimens that Andrews sent back to New York were primarily intended for exhibition rather than scientific purposes. As such, there were only two requirements for selection: first, that the museum did not already possess the species, and second, that they be big.[54] Smaller whales simply did not qualify as organisms desirable for shipment back to New York. When a particular sperm came into Aikawa, Andrews told Allen, "It is a perfect specimen—an old male sixty feet long. I think I would wait a long, long time before I'd get either a larger or a better specimen."[55] "So far as I know," he wrote in another letter, "this is the largest of any specimens now in a Museum."[56] It may seem peculiar for a naturalist to select a specimen only because of its bigness, but what made whales interesting for Andrews was precisely their size. He viewed whales as game, and the bigger, the better. Moreover, he was thinking ahead to exhibiting these specimens at the American Museum. In order to create a sublime sense of awe and wonder in the museum spectator, Andrews sought out the most mammoth specimens possible.

The acquisition of large specimens was also tied into the culture of sportsmanship that imbued Andrews's practice as a naturalist. The story of Andrews at the harpoon cannon, mentioned at the start of this chapter, took place on this expedition. If Andrews spent his time here adding marine megafauna to his game list and collecting massive oceanic trophies for display, he was by no means a peculiar American. The idea that the ocean could be a geography

Roy Chapman Andrews posing in front of a sperm whale in Oshima, Japan, 1910. Andrews would take measurements of the whales on these shore stations. Some whales were cleaned, boxed, and sent back to the American Museum of Natural History in New York. Courtesy of the American Museum of Natural History.

of sport—specifically for anglers and hunters—began around the turn of the century with the development of deep-sea sport fishing. But the gaming of ocean fauna has a peculiar history. It sits on the fence that divides the cultures of anglers and hunters, prosaically set out in the conversation between angler, hunter, and falconer in Isaac Walton's classic *The Compleat Angler, or The Contemplative Man's Recreation* (1653).

The relative virtues of rod versus gun were the fodder of conversations between dissenting men in club smoking rooms in the first decade of the twentieth century. A day with the rod after trout or salmon promised solitude, communion with nature, rest, relaxation, and a clever matching of piscatorial instinct and human ingenuity—all necessary ingredients during a "busy man's holiday after a spell of brain fag." The hunt for wild game, in contrast, involved the vitality and strength of youth, steady nerves for a straight shoot, the percussive cacophony of a company of rifles, and a view of nature, one critic ruefully noted, that encompassed "nothing more inspiring than fallows, roots, or stubbles." Hunting and angling were, of course, united under the category of

A 1934 photograph of Roy Chapman Andrews in front of the skeleton of a right whale that has been transported and erected at the American Museum of Natural History. Courtesy of the American Museum of Natural History.

"sport," a leisure activity of wealthy and elite classes that pitted humans against nature. But each was differentiated by a set of distinctive values.[57]

Angling in nineteenth-century America was primarily a freshwater affair. But at the turn of the century, just as the cult of ocean swimming became a new fad, wealthy American sports fishers began plying the waters of the Caribbean and southern California for saltwater fish. This was not merely an extension of freshwater angling interests; when saltwater fishers entered the oceans for yellowfin tuna, broadbill swordfish, and marlin, they did so as both anglers and hunters from America's wealthy elite. As a consequence, the prey of the deep-sea fisher—or rather, the deep-sea hunter—became "game."

Sea fishing first became popular in south Florida and southern California. On the Gulf Coast of Florida, tarpon became an extremely popular game fish that, according to one fisherman, "thrills the most stolid of human participants to the tips of his toes, and to compare with it any kindred sport is a tiresome travesty."[58] Both coasts of south Florida became centers of tarpon enthusiasts just as the tuna was noticed in the waters near Avalon, California. A group of tuna fishers came together to form the Tuna Club of Catalina (1898), which throughout the first three decades of the twentieth century was the mecca of Pacific big game fishing. The sport thrived as new organisms were added to game lists: the broadbill swordfish, marlin, sailfish, striped bass, and bonefish, each new fish promising a set of distinctive characteristics that made them valid species for sport. Elite fishing organizations mushroomed throughout the sunbelt and beyond: the Long Key Fishing Camp, The Tyee Club of British Columbia, The Sailfish Club of Palm Beach, The Light Tackle Club in San Francisco, and the Miami Beach Rod and Reel Club.

Just as hunting clubs had established prizes and contests for biggest game shot, so too did deep-sea anglers compete for cherished immortality in having their names associated with catching the heaviest fish of their respective classes. In 1930, Earl Roman, a fishing editor for the *Miami Herald,* pointed to the need for anglers to associate and agree on rules, specifications, and ethics so that anglers from all around the world would be able to compete for these coveted records. The Salt Water Anglers of America, a national body dedicated to research, conservation, restoration, and, notably, battling pollution of the oceans, responded to this need. They also promoted sportsmanlike methods of angling; the association laid down rules, class sizes, equipment standards, and ethical guidelines that standardized the practice of big game fishing.[59]

The gaming of ocean fauna represented a move to extend America's frontier into new space, namely the oceans surrounding the continent. Like western game, ocean fauna promised a test of strength. The president of the Salt Water Anglers of America noted, "He of faint heart should not take up the sport of big game fishing, for the taking of large and stubborn sea monsters presents thrills and problems that call for good generalship and a hefty brand of stick-to-it-iveness."[60] Sea fishing provided a new place for the contest between civilization and the savage frontier. Moise N. Kaplan, one of the several authorities of south Florida sea fishing, claimed that "the marine assaulter . . . is equipped with inherited instinct, with crafty reasoning powers and appropriate perception. Gifted in making sudden and violent approach—contact with the enemy—silently, unobserved, he is able to harass and fatigue the defensive element while guarding and shielding himself."[61]

There is no greater evidence of this movement from western frontier to ocean, from hunting to deep-sea fishing, than the life trajectory of the great western novelist Zane Grey, also the holder of the yellowtail world record—one hundred eleven pounds on light tackle—for over ten years. In the first three decades of the twentieth century, Grey became one of the foremost mythologizers of the American western frontier. Grey's scores of novels and short stories were among the most widely read representations of the West. He introduced the beauty of the western landscape to eastern urbanites; his stories are full of struggles and battles between different human cultures. He focused on "cattle culture imbued with individualism, rustling, and justified violence." And he glorified the virtues of hunting game. The overall moral of his stories was to describe how remnants of frontier culture remained extant in the twentieth century.[62]

In the 1910s Grey could be found deep-sea fishing off Florida's coasts, and in 1924 he purchased a three-masted schooner, rechristened it the *Fisherman,* outfitted the vessel with ocean fishing gear, and sailed to the fertile waters off southern California. The chronicle of this expedition was published in his *Tales of Fishing Virgin Seas* (1925), a book that codified his status as America's most prominent spokesperson for ocean game fishing—a title that would pass to Ernest Hemingway in the 1930s. For Grey, the Pacific was a virgin land, a frontier of adventure, sport, and abundant resources. The frontier themes that characterize his western fiction can all be found in his experiences with tuna, yellowtail, and swordfish. And Grey was just one of the many hunters and fishers who had cut their teeth in the frontier West before exchanging horse for ship, gun for rod, and elk for marlin.[63]

Grey made even a more formidable contribution to Andrews's world of metropolitan-based natural history display. In 1928 the American Museum opened the doors to its new hall, Fishes of the World, an exhibit that would, according to naturalist William Gregory, "keep our visitor fascinated with the wonders of the fish world on the trip around the hall." The climax of the entire exhibition was the collection of big game fishes. The background display of the sailfish group portrayed the rocky islands of Cape San Lucas and featured a battle between a nine-foot sailfish breaching the ocean's surface and a deep-sea angler in a nearby boat who "pits his quick hand and unflinching will against the plunging weight of the maddened fish." The entire north wall of the exhibition displayed the mounted specimens of ocean sunfish, tunas, marlins, and swordfishes—all the trophies of Mr. Zane Grey, the well-known "Nimrod of the Seas." Indeed, the worlds of hunting and natural history—of mounted trophy and preserved specimen—converged in both field and museum.[64]

Clearly, to call Andrews an ocean fisherman would be foolhardy. His game was the whale, and his weapons were the camera and the harpoon gun. Nevertheless, Andrews was part of a larger project of transforming the ocean into a geography of sport, an endeavor in which ocean fauna became game for America's leisure class. Oceanic life was taking on new meanings in these early decades, and Andrews became both a participant and an architect of the transformation of ocean fauna into game—an event that was driven by a certain frontier anxiety, a desire to recreate America's western frontier in new parts of the world.

Giants of the Mackerel Family displayed in the American Museum of Natural History's Hall of Fishes, 1938. Some of these specimens were donated by Zane Grey, who can be seen posing next to a fish in the photograph to the left. Courtesy of the American Museum of Natural History.

JAPANESE WHALE FISHERIES AND PLANKED WHALE STEAK

Always a keen observer of whaling operations and procedures, Andrews's work abroad helped to fashion a separate, but related, facet of his conservation ethic; the Oriental Whaling Company was also a meat butcher and canner that brought whale meat to Japanese markets. The Japanese people, according to Andrews, made much better economic use of whale products than their American and European counterparts, who wastefully converted whale meat into fertilizer. "A large whale in Japan is worth $4,000; for a whale in any other country in the world $1,000 would probably be its greatest value: then, with the present use of whales for guano and oil, there is a tremendous amount of waste."[65] Although American shore stations practiced marked efficiency over their nineteenth-century predecessors by not setting whale carcasses adrift, the Japanese people were even more efficient because they converted cetacean meat into human food, intestines into leather, and finback baleen into cigar cases, charcoal baskets, sandals "and other beautiful things created by their clever brains and skillful fingers from the material which in the hands of Western nations seems to be almost useless."[66]

More than cigar cases and sandals, it was as meat that Andrews saw the greatest potential for the efficient use of whales. But most Americans seemed to have difficulty reconciling the fact that an ocean-dwelling organism could have a flavor closer to venison and beef than cod.[67] Nevertheless, many boosters of the whale industry held that there was "a practically inexhaustible supply of whale in the Pacific off Puget Sound"—a vast trove of easily accessible resources that could be served to Americans on the cheap.[68] Beginning in 1911, Andrews began encouraging Americans to break with their prejudices against consuming whale meat. He was corresponding with the superintendent of Pacific fisheries, who informed Andrews that some stations were retrofitting their factories in order to ship canned and fresh meat to market. A move was afoot to do likewise with all Alaskan and British Columbian whale fisheries, but the effort would prove fruitless if Americans refused to buy the product. More than anything else, it was the meat shortage of the Great War that caused Andrews to see in this "venison of the sea" "a promise of abundance in a time of scarcity."[69]

During the war, several scientists at the America Museum were anxious to contribute to the national good by helping to alleviate wartime shortages of beef, poultry, and pork. Through the leadership of President Wilson, the Federal Food Administration mounted a campaign calling for conservation of these staples. Henry Fairfield Osborn called on Andrews's knowledge of whales to publicly demonstrate a potential alternative to American meat consumption, so Andrews assisted in organizing a luncheon at the Museum that featured whale hors d'oeuvres and planked whale steak *à la Vancouver.* In attendance along with Andrews were scientific and professional notables such as Osborn, Frederick Lucas, Charles Townsend, William Hornaday, Admiral Peary, and Caspar Whitney. The event was billed as a "conservation luncheon" demonstrating "the utility of whale meat as a substitute for beef." This group of New York scientific and business elite unanimously expressed the sentiment that whale meat was as "delicious a morsel as the most aesthetic of sophisticated palates could possibly yearn for."[70]

The politics of social class was a subtext of the event. Andrews had previously noted that whale meat was the staple of Japanese classes too impoverished to buy beef.[71] After consulting Andrews, Osborn stated that the great appeal of introducing whale meat into the national war diet was that it could be supplied annually at only twelve and a half cents per pound. The *New York Times* printed an editorial that highlighted the contradiction of an epicurean luncheon for elites that nominated the whale "as a candidate for the poor man's table." Despite the purported low price of whale meat, there was no guarantee that prices would remain low and that whales would become provender for the masses. The U.S. Fishery Service had previously introduced the tilefish with similar economic promises, only to have the price skyrocket. Whale meat seemed to have its greatest potential as a curious victual for elites. The *Times*, thumbing its nose at such social snobbery, reported that "Dr. Osborn's diners testified that whale had a rich venison taste, which suggests an accompaniment of full-fruited burgundy."[72] According to the editorial, the luncheon resembled the highfalutin annual dinner of the Explorers Club that would gather once a year at the Waldorf Astoria to hand out accolades and dine on such curiosities as a morsel of 250,000-year-old hairy mammoth meat.[73] Some eight months later, the *Times* wrote a follow-up piece reporting that whale meat was selling at twenty-five cents per pound and that, although the sale of whale meat started in the poorer quarters of the city, it "ended in the uptown district, which considers itself to have higher standards of living than downtown. To be plain about it, the richer classes bought the 25 cent meat away from those who needed it more."[74]

The gender implications of whale consumption were more subtle than the class dynamics. The Great War gave rise to a feminist-pacifist critique that linked the butchery of animals with the brutality of war—both being the province of men. Cookbooks like *Meatless and Wheatless* (1917) and *The Golden Rule Cookbook* (1916) at the same time advocated vegetarianism during a time of war shortage and criticized Americans' aggression toward their human enemies as well as toward cows, pigs, and chickens. A full-blown modern vegetarian movement thus simultaneously called into question the male desire to hunt for food and kill in war.[75] It is not insignificant that Andrews had characterized much of his cetacean work as a manly hunt and many of his specimens as game. Meat was also believed to be crucial to the diet of America's fighting men. Admiral Peary made the link between whale meat and soldiers in a recommendation that the army use jerked whale meat in soldier rations. "It is the only prepared meat food with which I am acquainted," Peary wrote to General Geothals, "that men at hard work in the field at low temperatures can eat twice or three times a day . . . and keep in the best of health."[76] In response to vegetarians who called for the abandonment of male/meat dominance and Wilson who encouraged Americans to go meatless as a sacrifice to soldiers on the front, Andrews and Peary attempted to secure the place of whale meat in the American diet of both soldiers and civilians. Andrews's activities seem to be motivated, at least partially, by an attempt to preserve the virile traits threatened in the modern era.

Andrews never made whale meat a staple of his diet; nor did other AMNH naturalists or the American public. Although whale meat enjoyed continued boosterism and even some success on the West Coast, whale canneries quickly fell on hard times.[77] The important point is that in Andrews's hands, the whale was transformed into another commercial product. His actions fit squarely into a Progressive conservationism that demanded the wise use and scientific management of natural resources. Andrews took this message to heart; but what is more intriguing is that his contact with Japan served as a model for his American-style conservationism. Progressive conservation was not always a homegrown American movement; it was time and again forged in the field where the naturalist engaged the nonwestern world.[78]

THE KOREAN EXPEDITION AND THE MODERNIZATION OF CETOLOGY

While in Japan, Andrews had heard reports that Korean whaling stations were taking gray whales, a species that was then believed to be extinct in the eastern Pacific along the California coast. He wished to collect data and secure a number of specimens of the gray and also the many killer whales that were a staple harvest of the Korean fisheries. His second goal had nothing to do with whales; Andrews, who was always searching for those few remaining blank spots on the map, decided to explore portions of northwest Korea.

Again, officials of the Oriental Whaling Company, then in charge of Korea's whale fisheries, brokered the Korean expedition. In Ulsan, he spent some of his time on board whaling vessels taking photographs and making observations of the natural histories of gray and killer whales; but he mostly stayed ashore and waited for the whales to arrive at the stations, where he took his measurements and prepared his specimens. Andrews believed that such factories presented naturalists with a unique opportunity to secure large amounts of data in a brief period of time. The natural history of cetaceans could be most efficiently investigated by working hand in hand with modern whaling operations.[79]

He clearly thought that he was practicing a thoroughly modern natural history that had many advantages over the work of nineteenth-century naturalists. Previous naturalists relied on the occasional beached whale, which had often deteriorated before adequate descriptions could be made. As a result, naturalists made only a handful of observations, and those observations—especially variations in color—were often tainted by changes that occur shortly after death. The result of such a practice was a needless multiplication in taxonomic names. Nineteenth-century naturalists aboard whaling vessels were also beset with the problem of making accurate measurements on whales that were immediately "cut in" as soon as they were attached to the hull. Until shore whaling began, accurate photographs, notes, measurements, and descriptions were rare and unsatisfactory; anatomical work was out of the question. In contrast, "a naturalist who is fortunate enough to remain for some time at one of the shore-stations has before him wonderful opportunities."[80] One advantage was that a carcass on the slip was easy to measure, photograph, and describe. Another was that a great number of whales of a single species quickly paraded past the naturalist observing the daily catch; the sheer volume of whales facilitated the study of individual variation. As American factory owners embraced the mantra of Frederick Taylor's principles of scientific management, so too did Andrews view his own brand of natural history as an efficiently practiced science made possible through the technological apparatus of the modern whaling industry.[81]

A plethora of data was especially helpful for investigating evolutionary questions pertaining to cetaceans. Henry Fairfield Osborn's influence on the museum's department of vertebrate paleontology, and gradually over the entire museum itself, led many of its naturalists to embrace evolutionary projects. Andrews was no exception. When he first set out to make Pacific whales the object of his work, he was challenging the conventional wisdom that whales were cosmopolitan in nature—that is, that finbacks and right whales of the Atlantic were the same species of corresponding organisms in the Pacific.[82] Osborn's theory of radiational adaptation suggested that the wide dispersion of cetaceans would lead to significant variation and speciation. It was therefore a logical conclusion that organisms as widely dispersed as whales could have differentiated into separate species. By using Frederick True's study on Atlantic Cetacea as his model, Andrews wanted to author the definitive monograph on Pacific whales. The question of whether or not Atlantic and Pacific species were different would be solved through the careful analysis of the data from as many specimens as possible. This is where the value of the shore stations was most evident.

Although Andrews took photographs and wrote descriptions of the daily catch in all the stations he had visited, it was only on his final Korean expedition that the data would amount to a formal publication. He never produced the comprehensive study of Pacific Cetacea, perhaps because he came to realize that whales were, in fact, more or less cosmopolitan organisms.[83] Andrews claimed that continued fieldwork and many interruptions delayed the assembling of a single text. He did, however, write natural histories of two species, the California gray and the sei whales. His work on the gray was based entirely on his fieldwork in Korea, and it is here that we see how a modern and efficient practice of natural history translated into a scientific treatise.

Despite Andrews's talk of a modernized natural history, the several charts and tables that were the final result of this method provided little information for an evolutionary discussion. Most of his evolutionary conclusions were based on a single museum specimen. So it is difficult to understand why Andrews would go out of his way to call attention to a modernized natural history. Years later, while leading the Central Asiatic Expedition, Andrews wrote prolifically on the "modern business of exploration."[84] It seems clear that he was then using this rhetorical device—that is, his work being conducted as a modern business—to appeal to New York's financial elite, who were underwriting the expedition. Perhaps Andrews was using a similar strategy with his inquiry into whales.[85] Whatever the cause, Andrews was clearly clothing himself in a culture of scientific efficiency that likely won over a number of admirers.

In any case, *Monograph on Gray Whales* (1915), an AMNH publication, was perhaps his most important direct contribution to natural historical knowledge. The monograph was received as an exceptional piece of natural history. Indeed, Andrews was on his way to becoming one of America's leading naturalists of Cetacea. Henry Fairfield Osborn thought as much and wrote, "Through the lamented death of Doctor True, you have suddenly become the principal worker upon Cetaceans in the United States. It is a great opportunity and I feel that you will rise to it."[86] He did not. Always fashioning his own fate, Andrews's interests quickly turned as a result of the second half of his Korean expedition. His exploration of the Korean White Mountains spurred his wanderlust and his desire to conduct research in little-known regions. In November 1915, Andrews decided to specialize in the study of Asiatic zoology, thus setting into motion a chain of events that would eventually bring him fame as a Gobi explorer.[87]

In the 1930s Andrews turned his administrative skills away from fieldwork and toward the institutions that had supported him over the years. He took the reins of the Explorers Club and successfully navigated it through a period of economic strain brought on by the Depression. In 1934 he became director of the American Museum, where his leadership was marred by criticisms of a lack of foresight and claims of careerism. In 1941 he retired to a small ranch in suburban Connecticut, where he and his new wife worked on various autobiographical accounts of his life. His work on Pacific cetaceans did not endure for long, and his two important monographs are rarely cited in cetacean literature. As early as 1927, Robert Cushman Murphy was considered to be the AMNH's main expert on cetaceans. All that remained of Andrews's legacy to oceanic natural history were several whale skeletons and the life-size model of the blue whale that hung in the museum's Hall of Ocean Life.

Andrews's exploration of Pacific whales provides us with a window onto the landscape of postfrontier America and its surrounding oceans. As the eyes of a new American empire began to gaze beyond the West to other regions across the globe, the ocean itself began to absorb the cultural meanings and uses of an American frontier. Andrews moved into the North Pacific as a representative of an elite class of urban hunters who were routinely preoccupied with the sport of hunting, the accumulation of trophies, the efficient conservation of natural resources, the preservation of wilderness tracts of land, and the business of exploring new geographies with a naturalist's and imperialist's sensibilities. This hunting ideology informed his practice as a naturalist as well as the meanings he gave to whales and the ocean. His popular representations of whales and the ocean—in the forms of museum displays, articles, books, public lectures, and luncheons—also reflected that ideology. Moreover, his history is not entirely peculiar. Other naturalists and sportsmen were heading to the ocean with the same purpose of finding resources—to either exploit or conserve—and game for the hunt. But not all Americans shared this view of the ocean. Some naturalists looked at the modern ocean and were filled with nostalgia; they looked at the human use of the ocean and saw a familiar frontier process in action. They recalled the lessons gleaned from a century's worth of pioneer ax swinging, and they made an ardent call for the conservation of the ocean and its resources.

CHAPTER TWO

Robert Cushman Murphy and the Natural History of Ocean Islands

Conservation and the Myth of the Inexhaustible Ocean Frontier

On October 12, 1912, approximately 820 miles due south of the Cape Verde Islands, a maturing bull sperm was foraging for food; strong winds and a torrent of rain whipped the ocean to a frenzy. He was about ten years old and had spent his life roaming the Atlantic's equatorial waters with a pod of about ten to twenty cows and calves. They all traveled together, at times joining larger gatherings and then splintering off. At other times, a mature male frequented the pod and mated with the females, then left for other feeding grounds in higher latitudes. This day was like any other, with the young sperm leaving the pod with upturned flukes and traveling to a depth that was rich in fish and squid. The sun, though shrouded by clouds, had just risen over the eastern ocean, and the scattering layer—an extraordinary community of concentrated marine life—was descending, almost like the horizontal filter on a French coffee press. There he foraged for about twenty minutes, as his body automatically went into a physiological state that conserved oxygen. An ingenious evolutionary mechanism enabled him to emit strong pulses of sound from a circular section of his forehead powerful enough to stun squid and fish that would be easily sucked past his functionless teeth and into his gaping mouth. Oxygen depleted, the bull climbed to the life-sustaining atmosphere above, now using his echolocation capabilities to find the rest of the pod. As he leveled off, preparing to break the ocean surface, an iron harpoon entered his side. The bull leapt forward and breached the surface. His flukes grazed the keel of the wooden whaling boat from which the harpoon was thrown. If the bull had had any historical memory (in human terms), he may have thought it a curious irony that he had been plucked by a hand-thrown harpoon—this during an age of coal-powered steamships and harpoon cannons.

About thirty minutes before, the wooden whaling boat had been lowered into the ocean from the deck of the *Daisy,* one of the few remaining classic whaling brigs still engaged in the sperm whale fishery. The *Daisy* had just concluded its eastbound zig across the Atlantic and was riding the consistent trades on its zag to the southwest, bound for the elephant seal–rich island of South Georgia. It was very early in the morning when the masthead's cry announced the presence of two sperm whale pods to windward. The crew snapped into action, and within a few minutes, two whaling boats were making steady progress when the young bull sperm—still slightly submerged—crossed the bow of one of the thirty-foot skiffs. Emiliano Ramos, a twenty-two-year-old boat steerer from Sao Nicolau, rose from the bow, straightened his five-foot, eight-inch frame, and planted the iron harpoon into the animal's right flank, causing it to jump forward into a veil of mist. The crew hauled in the line, and through the morning rain, the bull was seen wallowing with the rest of the pod. The whalers continued to bring in the line, and with wood to black skin, one of the mates plunged a lance into the sperm's lungs. The bull breached and then sounded with extraordinary energy, and it was at that time that the whalers realized that they had made fast to a "forty-barrel bull," a whale notorious for its lusty spirit.

The small crew spent nine hours battling for their forty barrels of oil—a familiar game of Nantucket sleigh rides, paying line out and hauling it back in. One member of the crew, Robert Cushman Murphy, was a greenhorn to this activity; indeed, this was his first time manning an oar in a whaleboat. Murphy, a naturalist from New York, was under no obligation to engage in such labor; like Charles Darwin on the *Beagle,* he was merely hitching a ride aboard a ship to investigate the mysteries of nature. But Murphy was more than a naturalist; he reveled in every opportunity to become a functioning part of the living and breathing *Daisy* that, in his mind, was an anachronism in the era of modern whaling. To the rest of the mostly Caribbean whalers aboard the skiff, flailing away with gusto and enthusiasm, the whale—and the whale chase—were objects of desire. We can locate that desire in each individual's carefully apportioned lay of the haul, or perhaps in the fleeting approbation of the *Daisy*'s captain. Murphy, however, was receiving a symbolic lay of 1/200 the ship's take; and he, like Darwin, spent much of his time aft the main mast, a site of respect and honor. And so, while engaged in this epic battle of man against nature, Murphy confessed to a "certain sympathy with the enemy." With a characteristic prescience that outstripped the state of cetalogical research of the time, he further noted that the behavior of whales "required good mammalian brains. Whatever dim ideas were in them were quite unfishlike." Comparing them with the sea lions that were just learning to toss balls in zoological gardens in

America and abroad, he accurately predicted "that a whale's predilection for becoming chummy and companionable might astonish the world."[1]

If Roy Chapman Andrews represented the forward movement of a modern technocratic business elite into an ocean full of manageable commodities and game, Robert Cushman Murphy moved into the ocean as a historian—a historian who was passionately ambivalent about the progress of modern civilization. He was a historian of the naturalist sort, devoting his life to the time-honored practice of observing nature, compiling life histories of various aquatic organisms, and fitting them into the context of their natural surroundings. But he was also one of the many urban industrial professionals of Progressive America who felt quite uneasy with the quick changes of the modern world. Affiliated with this antimodernism was a kind of criticism, and other times outright hostility, toward a modern world defined by technology and heedless development. This sentiment was a vestige of a Romantic tradition that swept modern environmental thought beginning with the Transcendentalists and perhaps best represented by John Burroughs in Murphy's day. The criticism, however, was not doled out solely to the world of cities and sprawling suburbs. His antimodernism also framed a certain antireductionism—a desire to see connections between organisms, or rather, to view nature as a whole instead of an assemblage of parts.

In turn-of-the-century America, natural history was part of a tradition that was fast becoming somewhat irrelevant in the modern world of laboratory-oriented technoscience. In contrast, Murphy characterized himself as a "simon-pure" naturalist, and he believed that something was lost when complicated natural phenomena were reduced to chemical and physical abstractions. If the modern experimental scientist wanted to clean the slate and build a new system of natural knowledge, natural historians like Murphy thought of themselves as just the latest bearers of a torch that had been carried by the great luminaries of natural history. A detailed knowledge of Pliny, Linnaeus, and Darwin was as important as knowing your way around a specimen workroom. Murphy came of age during the early days of ecology, a growing discipline at the turn of the century that some commentators found different from "natural history" only in name. "'Ecology' is erudite and profound," noted the great American naturalist Marston Bates, "while 'natural history' is popular and superficial. Though, as far as I can see, both labels apply to just about the same package of goods."[2] Murphy's particular brand of ecology emphasized geographical and evolutionary questions—practices of science that were largely developed on land, and at times, in America's frontier West.

Murphy's antimodernism, rooted in his love for both nature and history, did not preclude a desire to use science for addressing the thorny conservation and environmental issues that were so prevalent in early twentieth-century America. Indeed, his American Museum colleague, Dick Pough, once remarked that "conservation is little more than applied natural history."[3] Murphy couldn't agree more, and he consequently became one of New York's prominent conservation leaders, always banging his drum to the beat of developing the practice of "conservation as scientific forecast." His conservation ideas were primarily influenced by the work of George Perkins Marsh, whose pathbreaking book, *Man and Nature* (1864), criticized the heedless rush of advanced societies that put profit and gain ahead of wise, and economically sustainable, resource use. Indeed, as Murphy's conservation ethic evolved, it became more pointed and to a certain extent Americanized in a way that evaded Marsh's scope. The key problem was not just big societies outstripping limited resources, as Marsh would have it; rather, the fountainhead of America's problems, Murphy believed, was the dangerous frontier myth of inexhaustible resources.

This is what became of Murphy's ocean, especially the islands and seabirds that became his primary subjects of interest. His thoughts on the forty-barrel bull were representative of his critique of the modern use of natural resources; his confessed sympathy for the sperm whale extended to the guano birds of Peru, the anchovetas of the Humboldt Current, the endemic life forms of New Zealand, and the wildlife of his native Long Island. The puzzling fact of modern American history is the resiliency of a "free lunch" idealogy—the idea that a purported trove of resources lay just beyond the civilized world. The ocean was as susceptible to this myth as was the American West and, later, outer space. Murphy's movements across the oceans, his scientific research, and his conservation efforts all demonstrate a sustained critique of that fundamental axiom. Speaking at a luncheon of the Garden Club of America, he noted that "the idea behind the new term 'proper land use' must, of course, extend its meaning to the sea. The wealth of life in the Sound and ocean, as it was described by our ancestors, is almost incredible reading today."[4] Here is clear evidence of Murphy's forethought; today, monographs on the emptying of the oceans now issue forth with great regularity. But to stop here would be to miss a more important point. Murphy's critique of the uses of the ocean, even the tools he used to investigate the ocean, drew from a history of terrestrial exploitation, a history of exploration in the frontier West. To make sense of the problems, potentials, and very nature of the world's oceans, Murphy simply extended a terrestrially constructed template onto wetter geographies. In short, Murphy's criticisms were a constitutive part of, not a move against, the transformation of the ocean into America's new wilderness frontier. His first exploratory activities did not bring him to the American West, as had many of his predecessors working at the American Museum of Natural History in the late nineteenth century. Instead, he went to the ocean to undertake a peculiar voyage.

THE CLASSICAL EXPEDITION OF THE DAISY

With ancestry dating to the *Mayflower* and a tradition of New England provincialism, Murphy grew up in northern Long Island as the eldest of nine children in a middle-class Roman Catholic family (though Robert Murphy's own faith shifted to Unitarianism at some point during his college years). As a youngster he became an avid naturalist of his local environs, then still an island abounding with wetlands, marshes, and forests. But like Andrews, it was the majestic spectacle of New York City's natural history museums that drove his vocational ambitions. He studied zoology at Brown and excelled. In 1911 he delivered a valedictory speech that no doubt led to some eye-rolling among his classmates, but it was quintessential Murphy. "The scientific movement," Murphy told his audience, "carries with it certain results which have by no means been hailed with unanimous acclaim, and which many thinkers consider unfortunate and even unsound." He disparagingly spoke of the declining status of the humanities, and, more important, to the regrettable disuse of the classical languages. How could we come to a deeper comprehension of "the evolution of man's life and emotions" without a knowledge of Euripides, Homer, and Virgil in their classical tongue? An odd speech for a scientist, but not so for a naturalist.[5]

After graduation, Murphy was presented with an opportunity that would frame his life's work. Frederick Lucas, then director of the American Museum of Natural History, offered Murphy a position as ship naturalist aboard an Antarctic-bound whaling brig for the purpose of examining the natural history of the subantarctic island, South Georgia. Reluctant to leave his newly betrothed on a long expedition, Murphy declined. But his fiancée, Grace Barstow, responded to the proposition with unexpected enthusiasm. In Murphy's words, she claimed that "the projected voyage would serve the best possible launching of my career, and that we would be married immediately so as to have several months together before my departure." The two were wed in mid-February of 1912 and honeymooned in the Caribbean while Murphy waited to rendezvous with the *Daisy*.[6]

The classical and antimodernist impulse of natural history was not merely a rhetorical tool for Murphy. His voyage aboard the *Daisy* highlights how historical and classical overtones seeped into the very practice of natural history. While the *Albatross* steamed naturalists around the world and wealthy owners of powered yachts gave passage to marine biologists, the AMNH Sub-Antarctic Expedition of 1912–1913 was a voyage *into* history. Most oceanic vessels at this time were propelled by the power of screw and steam. The *Daisy* was a relic of the past. Technically a half-brig, the ship was some 384 gross tons, 123 feet in length, two-decked, framed with oak and chestnut, planked with yellow pine, and fastened with copper. Two square-rigged masts, headsails, and a quadrilateral gaff sail aft of the mizzenmast had caught the winds in both the north and south Atlantic ever since it was launched out of Setauket Harbor—coincidentally very close to where Murphy spent his childhood summers. In 1907 a New Bedford group purchased and refitted the *Daisy* for whaling operations. Benjamin D. Cleveland, who Murphy venerably referred to as the "Old Man," was a shareholder and captain of the brig. The Old Man was a relic in his own right, a sailor of New Bedford sperm whaling stock who looked gloomily on the mechanization of whaling practices. One New York critic gave a perhaps characteristic description of the dying "old race of whaling Captains. . . . Lord! How they hated a steamer! Their occupation was dwindling. Whales got scarce. Mineral and gal oils crowded out whale oil. Rubber and steel usurped the place of whalebone."[7]

But why would a modern naturalist want to travel along on a slow-moving ship? For Murphy, slowness was a virtue. Murphy's presence on the *Daisy* was the result of a public dispute between Frederick Lucas and a Norwegian whaler, J. A. Morch, over the abundance of whales in the Atlantic. Lucas's thoughts on the "passing of the whale," which were so influential to Andrews's thinking, were not a matter of general consensus. Lucas's call for restrictions on whaling spurred Morch to write a series of published letters that pointed to the huge populations of whales that continued to inhabit the oceans, especially in the South Atlantic. Lucas agreed, but he thought that the relocation of Norwegian, British, and even American whaling to those waters would result, as in the Newfoundland case, in a decline in the whale population. Both were in agreement that a complete biological survey of oceanic life south of the fortieth parallel was much in need. Morch suggested that instead of putting scientific personnel on modern steamers, the most "expedient way would be the combination of a sailing, whaling and sealing expedition combined with thorough scientific research."[8] In 1911, Lucas approached Captain Benjamin Cleveland and secured a berth for a naturalist aboard the *Daisy*. Lucas offered the opportunity to Roy Chapman Andrews, who declined, and so the expedition went to Murphy.[9]

While lending a hand planning an impromptu wedding, Murphy made preparations for a yearlong expedition.[10] He gathered together his scientific supplies—shotguns, ammunition, alcohol, tools for preserving specimens, a Graflex camera, and thick binders of paper. He prepared a small library, including Dante's *Divina Commedia*, Bunyan's *Pilgrim's Progress*, Horace's *Carmina*, an *Oxford Shakespeare*, and Melville's *Moby-Dick*. Among the books devoted to natural history were Joseph Banks's journal of Captain Cook's first voyage on the *HMS Endeavour*, Darwin's *Voyage of the Beagle*, and Moseley's *Challenger Narrative*, the latter being one of the standard sources in modern oceanography. Grace had asked friends and family to write letters in advance—enough, when combined with her own letters and presents, to fill a gunnysack. Each item was dated so that Murphy would have at least some contact with home through the entire journey.

It is difficult for the contemporary reader to appreciate the meaning and purpose of such an expedition for a young naturalist with only a baccalaureate degree. He would later earn his Ph.D. from Columbia and become a prominent ornithologist at the American Museum of Natural History. But his graduate work—indeed, his entire life as a naturalist—began with the experience of the ocean. Murphy's future vocation remained wide open at this point in his life; his specialty would be those phenomena that piqued his interest the most. But as far as his work aboard the *Daisy* was concerned, he was a general naturalist with a fond interest in birds.

Harking back to a nineteenth-century mainstay of natural history, Murphy's daily practice consisted of the construction of the life histories of organisms. Life histories outlined all aspects of individual species, including morphology, life cycles, distribution, habits, behaviors, and connections with the past. Observing life cycles occupied the lion's share of Murphy's time: breeding behavior, nesting, and brooding. He was also interested in animals' mechanisms of locomotion, especially how their structure was suited to moving through particular environments, and so he busied himself shooting specimens from the sky to both record and preserve the organisms' structures. Sometimes he would trail a line of baited hooks from the stern. When the unsuspecting seabirds seized the bait and took flight, Murphy, and the crew he often enlisted to aid in his collecting, reeled in specimens, almost as if bringing in an animated kite. The captain ordered his crew to bring anything of interest to Murphy's attention, and so on several occasions, he was awakened from his berth by a masthead who had spied a curious whale or bird. He was even presented with an albatross that a crew member had shot from the sky and a large squid that a lanced sperm whale had regurgitated. From time to time, he left the confines of the *Daisy* and boarded a small dory, which Murphy christened *Grace Emile,* with notebook and rifle in search of specimens. More than anything else, he was intrigued by seabirds, especially those belonging to the order of Tubinares—petrels, albatrosses, and terns—that spent almost their entire lives as creatures of the ocean. Though Murphy's proclivities already leaned toward ornithology, he examined as many vertebrates as possible, including island rats, elephant seals, and cetaceans.

Murphy fully concurred with Lucas that the declining populations of the oceans' whales was a direct result of the modernization of the whaling fishery, and the *Daisy* expedition gave him a firsthand glimpse of the process. Nowhere was this more clear than at South Georgia. The island rises austerely out of the

Robert Cushman Murphy's photograph of the *Daisy* during a flensing operation. A portion of the sperm whale is being hoisted to the try pots. By 1911, when this photograph was taken, this method of whaling was rare. Courtesy of the New Bedford Whaling Museum.

frigid southern oceans some twelve hundred miles east of Cape Horn. Murphy described it as "one of the broken ring of mountainous, bleak, and treeless islands around the southern end of the earth."[11] It was the native home to a few species of grasses and flowering plants, and it was also the breeding ground for fur seals, sea elephants, albatrosses, and other seabirds. South Georgia was an island shaped by glacial forces and a blustery ocean; one could understand why James Cook, having visited the island in 1775, would call it "not worth the discovery." It was an island that loomed large in the journals of exploration that Murphy was reading. The triumphs and tragedies of these early explorers were recorded on the maps of the island: Possession Bay, Royal Bay, Discovery Bay, and Cape Disappointment. Although it was a meager island in the estimation of early explorers, South Georgia became a popular site for Yankee fur sealing in the mid-nineteenth century. After the demise of fur sealing, in Murphy's words, "the sea-elephants were forced to pay the costs of the ruthless voyagers." But in 1912, the whales of southern oceans, more than any other animal, brought Europeans and Americans to the island. As whale populations declined in northern waters, the result of the mechanization of the fleet, a large segment of the industry had redeployed to Antarctic waters. Beginning in 1904, South Georgia became a key nexus in the "exploitation of the southern oceans" with whaling slips erected in glacial fjords all along the western and northern coasts.[12] In fact, several whaling companies had established sizable outposts with many of the luxuries of urban living.

The *Daisy* arrived on November 24 and set anchor in Cumberland Bay on the eastern shore; the ship would cruise the coast for three months while its crew busied itself with the commerce of extracting oil from the island's elephant seal inhabitants. Murphy went ashore on the first boat and was immediately amazed by a beach "lined for miles with the bones of whales, mostly humpbacks. Spinal columns, loose vertebrae, ribs, and jaws were piled in heaps and bulwarks along the waterline and it was easy to count a hundred huge skulls within a stone's throw."[13] While the crew engaged in the bloody venture of the elephant seal hunt, Murphy built a temporary workshop on a tiny rock-rimmed cove in the Bay of Isles on the northern shore. In a well braced tent, he was at least partially sheltered from violent squalls as he proceeded to map, collect, preserve, and record the natural history of this little-known region.

Murphy was never a hermit; he befriended several of the sailors staying in Cumberland Bay who gave him the opportunity to spend a long day aboard a modern whaling vessel, the *Fortuna,* the first steam vessel to hunt whales in southern waters. The whalers had quickly realized that whales existed in staggering densities at an oceanic eddy that lay some thirty-five to forty miles off the northwest coast of the island. The spot was the point of convergence of two cold Antarctic currents from the southwest, and Murphy was beginning to appreciate how these icy currents provided the habitat for enormous quantities of krill and plankton. Such abundance, he was coming to understand, provided the all-important base of a food chain that supported large populations of seabirds and whales.

This was his first opportunity to witness the practice of modern steamer whaling. The power of wind and human muscle were made obsolete by coal-powered engines; the ropes for hauling sperms aboard were replaced with heavy gauged wires and winches; the need for whaling boats replaced by what Murphy termed "unsportsmanlike" explosive-tipped harpoon canons. The oil-extracting tryworks were moved from ship to either shore stations or to massive "floating factories." In short, modern whale fisheries were "conducted according to the most approved modern methods." Human muscle and sinew were replaced with the efficiency-minded technologies of the modern world.[14]

At one point, Murphy counted eleven other steamers at the happy hunting ground, and "the banging of harpoon guns was almost continuous."[15] The company to which the *Fortuna* belonged had processed sixty whales that day, and there were about six other companies in operation. So the species-destroying potential of the new practice was clearly evident. The facts were all too obvious: the total annual catch in the year 1912 was equal the sum paid for all the oil and bone brought home by New Bedford ships during the fifty most lucrative years of the old port's history. Writing a number of years later, Murphy noted that the new whalers had "killed the goose that laid the golden egg and only international conservation could limit and regulate the industry so that it may some day yield a sure and reasonable perpetual return."[16] Murphy, echoing a sentiment articulated some fifty years before by George Perkins Marsh, held that when human history supersedes the history of nature, the natural effects are often devastating, and conservation becomes the only mechanism for redressing the balance.[17] Action on the problem waited until the formation of the International Whaling Committee in the 1930s. In 1940 Murphy chronicled the spectacular growth of Antarctic whale fishing. Norwegian and American whaling stations had a similar history, but the opening of the Antarctic grounds lent a new impetus to the exploitation of wealth which, Murphy noted, seemed "inexhaustible." Even though the international community had taken notice by 1940, Murphy observed that "the number of whales being slain is at least fourfold what the oceans can endure on a long-term basis, yet the goal of reasonable, and hence perpetual, utilization seems farther off than ever."[18] His argument was reminiscent of Teddy Roosevelt's lobbying for the creation of the National Forest Service to conserve the timber resources of western territories.

In a sense, the destructiveness of modern whaling was not Murphy's story. He would later write only a single article, chronicling the role of naval might in early twentieth-century America, that dealt with the modern whale fishery. For more information on the subject, an interested reader would likely turn to the publications or presentations of Andrews. No, Murphy's story was almost psychologically linked to the *Daisy* and the operations of old-style Yankee whaling. Just as he was interested in describing the evolutionary histories of seabirds, so too did he most readily chronicle the human history of sperm whaling. As is so often the case, the naturalist is blown into the future facing backward.

To the captains and crews aboard the twenty steamers in Cumberland Bay, the *Daisy* was a throwback to a time when whaling was inefficient and labor intensive, much like the way modern automobile riders look at the Amish drivers of horse-drawn buggies. In contrast, the historic endeavor captivated Murphy; indeed, he thought that much of the charm, nobility, and grandeur of the voyage was a function of the *Daisy*'s antiquated practice. Just as Darwin's *Voyage of the Beagle* functioned as a kind of template for his work as a naturalist, so too did Melville's *Moby-Dick* serve as Murphy's guide to the sperm whaling activities of the *Daisy*.

Upon his return, Murphy became something of a minor historian of classic whaling. He contributed three articles to a magazine, *Sea Power*, that described and illustrated an industry whose days were most certainly numbered. What becomes clear in these descriptions is the massive amount of energy invested in classic sperm whaling: the hunt, the haul back to the ship, the cutting-in process in which the sperm whale was physically attached to the ship while the crew worked rapidly to lessen the take of the surrounding sharks, and the conversion of the beast into oil. The traditional practice of whaling required high quantities of energy and work channeled through both human muscle and ship—as distinct from the latent energy trapped in coal. Whalers came to know their subjects only through a lifetime of arduous work; whale oil became an artifact of whale blubber and human sweat. The subject would be the main purpose of Murphy's books chronicling the expedition, *Logbook for Grace* (1947) and *A Dead Whale, or a Stove Boat* (1967).

Murphy was by no means alone in his desire to preserve the history of the sperm whale fishery. Perhaps it was the change in the whaling industry, or more likely the pervasive technological and social transformations of modern America that caused many New Englanders to begin examining the history of whaling. Think of it this way: Melville's classic treatment of the industry was coolly received and relatively unread until the early decades of the twentieth century, when *Moby-Dick* was rediscovered as an American classic. Many literary historians believe that it took this much time for the merits of the novel to be recognized. But seen from a different light, the power and impact of the classic required a context in which many New Englanders were recreating their historical memory in an attempt to come to terms with the modern world. One writer referred to the economic and social decline of New England fishery towns as "the crumbling ruins of antiquity, which in some measure, tell the history of the decadence of industries that prospered in other years."[19] These new ruins gave rise to an almost palpable desire to preserve the past in published histories and museums, an effort that might be viewed as an attempt to prevent further decay that others might call "modernization."[20]

Murphy, too, played a modest role in the material display of classic whaling. The *Daisy* was recommissioned to transport foodstuffs during World War I (she actually sank when a leak allowed a cargo of dried beans to expand and tear the hull asunder). Before her leguminous demise, Murphy secured one of the whale boats fully equipped with oars, paddles, sail and spars, harpoons, lances, darting gun, line tubs and line, compass box, "and every other item, which would make her ready to lower for sperm whales." The American Museum of Natural History purchased the outfit from Murphy in 1928 and put it on display in the newly dedicated Hall of Ocean Life. In an action that Murphy regarded as a "serious error of judgment," the museum's exhibit committee later voted to have the boat stowed away in permanent storage.[21]

The American Museum justified the removal of the whaling boat from the Hall of Ocean Life by pointing to four murals of classic whaling painted in the lunettes on the north side of the hall that, in their view, adequately represented the historic industry. When the Hall of Ocean Life was being designed in 1927, Murphy coaxed the museum into commissioning John Benson, a highly regarded marine artist, for murals that would "present the romantic and beautiful side of whaling life, with the attendant dangers and the human courage displayed by the early American whalers, rather than the squalid, modern commercial side."[22] After much dispute over whether the function of the murals was to portray the habits of sperm whales or memorialize the whaling industry, the committee decided to give Benson the commission.

Most of the immense hall was a testament to the modernization of ocean travel. The *Arcturus* Exhibit portrayed the operations and specimens of William Beebe's 1925 expedition made possible through the loan of Henry Whiton's massive coal-powered cruiser. Roy Waldo Miner's Coral Reef Group (later updated to the still extant Pearl Divers Group) was the product of Templeton Crocker's loan of the *Zaca,* a powered schooner yacht. The habitat group for the northern elephant seal contained specimens from Charles Townsend's 1911 expedition to Guadeloupe aboard the U.S. government research vessel *Albatross.* And above them all were Andrews's whale skeletons, gifts from modern whaling companies. The hall was one of the largest in the world. At one hundred sixty by one hundred thirty feet and a painted blue sky some sixty feet above the floor, "the general appearance," Harold Anthony thought, "is somewhat suggestive of a cathedral."[23] Amid this sublime display of oceanic relics,

Clifford W. Ashley was among many artists and curators who, at the start of the twentieth century, were actively trying to preserve the history of New England whaling. *The Whaling Brig 'Daisy'* or *A Dead Whale or a Stove Boat*, c. 1920, oil on canvas. 25 × 20¼ in. Courtesy of the Long Island Museum of American Art, History & Carriages. Gift of Grace Murphy, 1973.

bankrolled by New York's wealthy elite and obtained through the modernization of ocean travel, Murphy's whaleboat enjoyed a short life. This technology of humble means thus became an artifact of natural history itself.

In 1936, Murphy beseeched Charles Davenport, director of the biological lab at Cold Spring Harbor, to provide a good home for the whale boat. Murphy's contribution became part of a larger push to create a museum "for the preservation and exhibition of antiquities and relics pertaining to the whaling industry and era."[24] In 1943 the museum was finally erected at Cold Spring Harbor. Murphy provided his services as a consultant and also presented a series of Graflex photos he had taken back in 1912. In 1963 Murphy ceded to the museum a small library of whaling books, logbooks, several pieces of scrimshaw, and various other ship gear that he had collected throughout his life. The Whaling Museum was so indebted to Murphy's contributions that they placed an image of the *Daisy* on their letterhead.

Murphy's nostalgia reveals a certain regret over how America was interacting with the ocean in the early twentieth century. Upper- and middle-class Americans, as well as new immigrants to America, were fast becoming familiar with the ocean steamliner. Ironically, the more Americans traveled aboard ocean liners, the less they actually encountered the ocean. The modern ocean steamer, after all, was a form of transportation designed to help the passenger forget that they were traveling on the ocean, to create the impression that the passenger had never left the security and comforts of home (unless in steerage, of course). The new form of transportation produced a torrent of articles and books: manuals, guides, and commentaries on such topics as the best lines, the quickest ships, and the most luxurious accommodations. What is stunning about these publications is the absence of the ocean itself. Just as the railroad transformed people's sense of space and time, so too did the modern ocean liner attempt to make the oceanic environment all but irrelevant—up to the point, of course, when icy seawater crossed the boundary of the bulkhead.[25]

Something was lost in such a world. While traveling to America aboard the luxury cruiser *Tuscania* in 1923, Joseph Conrad penned a short analysis of the radical psychological changes that ensued as a result of the drastically shifting material technologies of ocean travel. As with many of his writings, Conrad thought back to a more innocent past, a past when people traveled under the machinery of sails. With the advent of steam propulsion, Conrad thought, "the whole psychology of sea travel is changed." The old-time traveler "had to become acclimatized to that moral atmosphere of ship life which he was fated to breathe for so many days." The modern traveler brings along all the trappings of civilized life and looks forward to the moment of quick disembarkment, "but the other lived the life of his ship, that sort of life which is not sustained on bread alone, but depends for its interest on enlarged sympathies and awakened perceptions of nature and men."[26] This change in the traveler's psychology, deeply rooted in the material practices of ocean travel, had a significant impact on the practice of Murphy's natural history. Roy Chapman Andrews traveled on steam-powered whaling vessels and called for the modernization of natural history; Murphy traveled on a ship from the past and relished the history of natural history.

Along with making Murphy a minor authority on both whales and classic whaling, the *Daisy* expedition was also influential in turning his attention to the ecological connections between seabirds and the oceans. As they traveled further south, these birds showed up in greater numbers. By the time the *Daisy* found safe harbor off South Georgia, Murphy had already amassed the collection of albatrosses, petrels, shearwaters, and storm petrels that would constitute the material for his master's thesis. Murphy was particularly fascinated with the order of Tubinares, which consists of albatrosses, giant fulmars, cape pigeons, whalebirds, and Antarctic petrels. He considered them an "archaic" order as indicated by "the joint evidence of their structural isolation, ontogeny, psychobiological reactions, distribution, and the fossil record," which demonstrated that the order had reached its radiational climax by the late Tertiary.[27] Also of note was their great range of interordinal size, which finds no parallel in other ordinal groups of birds. How could these birds, with such closely related evolutionary histories, range so dramatically in size and shape? To explain this variation among close relatives, Murphy endorsed the adaptive radiation theory of American museum naturalist Henry Fairfield Osborn. Finally, Murphy was overwhelmed by the order's huge population.

Although subantarctic waters were species poor, they were teeming with these birds. Murphy took a page out of Osborn's *Origin and the Evolution of Life* (1917) and pointed to the homogenous and archaic nature of the polar oceanic environment. The absence of denitrifying bacteria in frigid polar waters explains the overwhelming abundance of the nitrates that support marine algae on which zooplankton subsists, thus founding the base of the ecological chain that supports pelagic birds.[28] So in order to explain the existence of such an archaic order, Murphy pointed to the historical oceanic conditions that supported the great populations of oceanic birds. He never completed the book-long treatise that he wanted to write on the evolution of seabirds, but a general course of study was set in place. Murphy was becoming a chronicler of the human and natural histories of the ocean and of life.

THE GEOGRAPHY OF OCEAN CURRENTS AND PERUVIAN GUANO

Beginning in 1919 Murphy undertook a series of expeditions to the guano islands of Peru and Ecuador to further explore the relationship between ocean, birds, and humans. Despite their proximity to the equator, Murphy's interest in these islands was actually a continuation of the subantarctic expedition. For the same type of fauna he had examined aboard the *Daisy*—southern kelp gulls, white-breasted cormorants, diving petrels, penguins, and the southern sea lion—were also present in the confines of the Humboldt Current. Murphy had a number of specific objectives in mind on his 1919 expedition. First and foremost, he wanted to explain the extraordinary abundance of littoral animal life along the coast of Peru by studying oceanic conditions. His work in the subantarctic made clear the relationship between abundance and the cool phytoplankton-rich ocean that supported these birds. The Peruvian studies refined this base of knowledge by bringing his attention to how specific populations of birds were distributed within definite littoral biogeographies. A necessary corollary was to investigate "the life histories of the birds, mammals, and fishes, in order to understand and interpret their ecology." He also went to the islands to secure preserved specimens for museum collections and material for a permanent museum exhibit that would eventually be part of the Whitney Hall of Pacific Birds at the American Museum of Natural History—then his place of employment. He also took three hundred fifty pictures and recorded nine thousand feet of film that was later edited to become *The Bird Islands of Peru*. The expedition was not without its commercial purposes. He was to report back to his primary patron, the American import firm W. R. Grace and Co., on the possibility of acquiring access to guano and anchovy resources. Last, Francisco Ballen, a naturalist of the Guano Administration of Peru, invited Murphy to the area for the purpose of providing the advice and authority of an American scientist on the conservation of guano resources.[29]

Giving up the romance of the sail for the efficiency of the screw, Murphy steamed through the Caribbean at August's end of 1919, crossed through the canal, and skirted the Ecuadorian coast before landing in Pacasmayo and thence to Lima, where he passed the time reading about the tumultuous history of guano and the Peruvian economy. Shortly thereafter, he boarded the *Alcatraz* bound south for the Chinchas Islands, the first research station on his northward sweep along the full length of Peru's coast. The Chinchas struck him as beautiful and at the same time stark and tragic, a combination of geological, natural, and human history:

> Three tiny, bare, splintery granite rocks are these, evidently all one islet at some time in the distant past, yet small as they are, their name is known to the farthest ports in the world, and their share in making fortunes and in abetting tragedies, in debauching not only men but governments, has given them a place in history all out of proportion to their size. . . . The flat top of North Island, now occupied solely by a dense colony of guanays, is said to have been the site of a town of eight thousand inhabitants. No trace now remains.[30]

Just as he had done on South Georgia, Murphy spent much of his time in the field observing the feeding, nesting, mating, and brooding behavior of the seabirds. He carefully noted the amount of time they had left the island to feed in the ocean, and tried, to the best of his ability, to discover exactly where these animals foraged. He spent an equal amount of time aboard skiffs and schooners noting the temperature of the surface water in both longitudinal and latitudinal profiles, tracking the winds, measuring the rate of flow of the current with a Gurley current meter, and saving samples of water for analysis of biological and chemical content.[31] Murphy also used a tow net to sample the diatomaceous flora of the current along the seven-hundred-mile coast. It was well known that phytoplankton was the energy base of an ecological pyramid that sustained copepods and crustaceans that became the food for the anchovies on which the guano birds were completely dependent. There was a sophistication in these techniques that distinguished them from his earlier work. The science of oceanography was just then adopting new technologies to measure and quantify various oceanic phenomena, and Murphy used some of those techniques. But if he was becoming something of an oceanographer, Murphy was also practicing a kind of ecological oceanography. His overall goal, of course, was to link seabird life to oceanic conditions. Always the historian, Murphy also asked some peculiar questions—for instance, his mind mused over the history of the Humboldt Current itself.

Taking plankton samples from the ocean would give only a present profile, and Murphy was interested in the history of this current, the evolutionary tale of how it came to be that the guano birds were so intimately linked with this region. So in the same way a geologist turns to the strata of an outcropping to investigate geological history, Murphy took guano samples from various layers built up over time, especially the so-called fossil layers from the beds on Lobos de Afuera and Lobos de Tierra. These samples were sent back to the Carnegie Institute, where an analyst identified and quantified the diatom content. What came as a surprise to both the analyst, Albert Mann, and Murphy was the extraordinary uniformity in species type and number, both over time and space. Such uniformity was an important finding because it explained how the life of the Humboldt Current "must have become thus inflexibly adapted through the long duration of substantially invariable conditions." Furthermore, it explained how slight variations in the current—like the El Niño phenomenon—could have such a dramatic effect on all ocean life.[32]

The 1919 Peru island expedition marked Murphy's first rigorous entry into the fields of biogeography and—generally speaking—geography. Murphy was attempting to bring the tools of geography to describe the Humboldt Current. Of course, the heterogeneous and dynamic nature of the current had not escaped the attention of marine biologists and oceanographers. But the geography of oceanic phenomena—currents, convergences, upwellings, and the like—was still in need of more systematic investigation. Murphy wished to understand the movements, temperature profiles, chemical composition, and biology of an important current that had a long history of exploration.

Murphy explained why water became steadily warmer as it became further removed from the coast and at the same time remained constant (both on and off the coast) as the current drifted northward. He attributed this phenomenon to the fact that the Humboldt was not a surface current, like the Gulf Stream, but a deep-water current. A steady southern wind caused coastal surface water to push out to sea, being replaced by an upwelling of cold and phytoplankton-rich deep water. Perhaps his most important findings related to the El Niño countercurrent. This was before the era when El Niño was viewed as an oceanwide alteration in energy cycles; accordingly, Murphy's explanation was extremely local. He called into question the common belief that El Niño events were caused by the swinging of the Humboldt Current offshore and found that the actual cause for high water temperatures was the invasion of a warmer countercurrent from the north.[33]

These notes were printed in the *Geographical Review*, the publication of the American Geographical Society. Isaiah Bowman was then the editor of the journal, and the two became lifelong friends. No doubt Murphy viewed Bowman as the quintessential geographer, whose "pervading soul . . . sees the basic bearing of the frame of the earth and the physical as well as social ties of its inhabitants."[34] Indeed, Bowman's climatic and geographic work, *Andes of Southern Peru* (1916), played no small role in Murphy's description of the coast.[35] Murphy's name became increasingly present in notes, reviews, and articles of the *Geographical Review*. He also became involved in the society's affairs, serving on organizational and conservation committees. In 1943, the AGS presented Murphy with the Cullum Medal for his outstanding contributions in the field of geography.

Murphy's foray into geography is understandable, given his interest in the biogeographical distribution of ocean birds.[36] Nowhere was this more present than in Murphy's most significant contribution to natural history, *Oceanic Birds of South America* (1936).[37] The essential problem with accounts of South American seabirds, according to Murphy, was that previous naturalists had relied too heavily on birds taken out of their ocean environment—that is, most of the collecting had been done on the continent, not in the ocean where they lived most of their lives. Murphy's own work in the subantarctic and Peru, as well as the specimens that returned from the Sanford Whitney South Pacific Expedition, provided the data that enabled him to situate ocean birds into their pelagic and littoral contexts. He separated South American waters into fourteen regions and depicted an ocean that "abounds in invisible walls and hedges. . . . The majority of oceanic birds are bound as peons to their own specific types of surface water."[38] Breaking from the myth that ocean birds were free to roam at will, he showed how thirty-two species of ocean birds were inflexibly bound to historically specific life zones characterized by surface-water conditions.

The idea that the ocean can possess a physical geography was made explicit in 1855 when Matthew Fontaine Maury published his classic oceanographic text, *The Physical Geography of the Seas.* But Murphy's use of the life zone concept is more closely linked to the work of the director of the U.S. Biological Survey, C. Hart Merriam. Merriam was a physician trained in New York who had cut his teeth as a naturalist on the 1872 Hayden Survey of Yellowstone Park. Various other jobs as a government scientist gave him the training to land the position at the biological survey in 1885. The survey was intended to be a service to the American people, especially farmers, who were increasingly making a living from the westernmost landscape. His team of scientists made recommendations of where certain crops grew best; they were also the people responsible for eradicating "pest" species like jackrabbits and wolves.

Merriam's work linked the practical with the theoretical. He thought the best way to advise farmers was to shine the light of science on the vanishing frontier West. So he created a new geography of the United States, one not separated by political boundaries but rather by nature's properties, which Merriam used to define unique environmental assemblages. His life zones were generally a function of average temperatures, and each zone contained a characteristic grouping of animal and life forms. Each zone was represented by a separate color that was graphed onto a map of the United States. The Governmental Printing Office published the reports and distributed them to farmers, much in the same way the Weather Bureau disseminated forecasts to the American public. Such biological cartographies were not new to Merriam's time, but in contrast to his predecessors, he thought that such information was decidedly practical. In a way, Merriam's life zones of the United States bear a resemblance to John Wesley Powell's *Report on the Lands of the Arid Region of the United States* (1879) in that both men were attempting to rationalize westward expansion on the basis of "nature's ways," instead of the grid logic of the Land Ordinance and the Homestead Act. Both understood the limits of the American landscape, and criticized, in their own way, the myth of the inexhaustible frontier.

When Murphy constructed the life zones of South American seabirds, he was taking Merriam's concept and extending it into the ocean. And just as with Merriam, such an understanding had practical implications. Beginning in the mid-nineteenth century, Peruvian officials seized on the massive accumulations of guano as a natural resource to be mined and traded to foreign countries. This industry produced enormous short-term wealth, but the government quickly exhausted the islands' reserves and in the process almost decimated seabird populations. As a consequence, the Peruvian Guano Administration was founded in 1909 to resurrect Peru's defunct guano industry by applying biological principles of life cycles that promoted guano production. The administration therefore strictly enforced a protection policy that, in effect, transformed islands into bird sanctuaries.

Murphy highly esteemed the conservationist strategies of the Guano Administration. In a letter to Francisco Ballen, he reported that "the extraction of guano is, fortunately for the nation, no longer a mere matter of exploitation; it has become an industry."[39] Instead of considering guano as a resource to be mined, the administration considered it a commercial product dependent on the contingencies of the seabird population. Protecting these populations was the most important objective of administrators. So by calling Peru's guano extraction an industry, Murphy was highlighting the administration's efforts to conserve and produce a profitable natural resource by protecting the birds that sustained it. Where nineteenth-century Peruvians were bent on accumulating massive wealth in the short term, their twentieth-century counterparts kept long-term sustainability foremost on their minds. Murphy was equally impressed with the administration's ability to adjust harvesting policy to accommodate for wild swings in seabird populations, usually a result of the El Niño countercurrent. In short, guano-producing seabirds had become the friends of the Peruvians, and this friendship preserved, in Murphy's words, "the balance of nature" and at the same time provided fertilizer for Peruvian agriculture and profits for the nation.[40]

THE ENVIRONMENTAL HISTORY OF ISLANDS

Immediately after the conclusion of World War II, Murphy, who by then had been awarded many times for his research and had assumed more prominent positions and responsibilities at the American Museum, was invited by Lieutenant Colonel Thomas G. Thompson, director of the Chemical Warfare Installation of the Army Service Forces in the Pearl Islands, to conduct a general natural history of the fauna on San Jose Island.[41] The island was located near the Panama Canal Zone in the Columbian Bight, an oceanic environment that he had investigated in the late 1930s. Murphy was struck by the massive changes that the military had made to the island's geography: "The labor expended here is simply unimaginable. Highways were cut and bulldozed, enormous gradings and fills completed, and the roads were surfaced and rolled with stone quarried and crushed on San Jose." But more than new roads sprawling over the landscape, Murphy feared the "heavy destruction of life all over the island by the war experiments." This suspicion was confirmed when he deduced the reason for the alarming paucity of bird life on the island. A naturalist who had explored the island before Murphy's visit went into the forest after a cyanogen chloride "shoot" and filled several gunnysacks with unmarked birds killed by the gas. Many of the birds on San Jose had met a similar fate. During the war in the Pacific, the U.S. military left a similar path of development and destruction throughout Oceania. The end of the war only promised further development. Murphy's Pearl Island expedition thus made clear that the militarization of Pacific islands could result in massive damage to these fragile environments. One of the consequences of this realization was his advocacy for conservation in the Pacific, a simple idea that was embodied in one of Murphy's most significant and enduring exhibits at the American Museum of Natural History.[42]

Patrons who had visited the American Museum of Natural History after 1953 would begin their tour outside by viewing the bronze statue of Teddy Roosevelt perched high on his horse with two indigenous guides at either side. Passing between the fifty-foot classical columns, they were greeted by three mounted *Barosaurus lentus* skeletons from the Jurassic, posed as if in violent battle. Carved into the granite on the front and back sides of the Roosevelt Memorial Hall were four quotations from Roosevelt that instilled messages about duty, honor, nationalism, masculinity, and the curing powers of nature. These imposing first sights were layered with meanings that reveal structures of power that continuously reestablish early twentieth-century assumptions of race, class, gender, culture, and nature. If visitors continued westward into Carl Akley's Hall of Africa, stories of race, class, and gender would be on display in the history and structure of the dioramas. Now, if they took a right turn and headed into the Whitney Hall of Pacific Birds, a far less imposing, but no less impressive, spectacle entered the field of vision. Positioned in the middle of the hall, they would be standing underneath a vaulted blue sky in what the curators of the exhibit believed was the center of the Pacific Ocean. As visitors moved from exhibit to exhibit, they would be traveling similarly along the map in such a way as to outline how the dynamics of ocean and climate vary geographically.[43] In short, the entire hall was a microcosm of the Pacific, divided into representative life zones.

At first glance, this seems to be a harmonious paradise—an Eden—just as the Pacific should be, according to Gauguin. Every diorama tells a story of a community of birds and the ocean that is painted into almost every background. They are lighthearted natural histories about family and courtship. They are also historical stories. The tooth-billed pigeon was a conspicuous representative of the Somoan group because it had "no very near relatives and is probably a relic of a branch of the pigeon group that has long since died out."[44] A model of the extinct New Zealand moa, completed in 1953, marks a diorama as a representation of something that once was, and no longer is. These representations of

The domed ceiling of the Whitney Hall of Oceanic Birds at the American Museum of Natural History, 1953. The exhibit was designed to make a visitor feel as if they were in the middle of the Pacific and birding along the Pacific Rim. Courtesy of the American Museum of Natural History.

nature are not about strength or virility or conquest. They calm the spirit and soothe the soul. They speak of a prelapsarian paradise, an Eden not only before the Fall, but a paradise untrammeled by the destructive feet of western *Homo sapiens.* Above all, these exhibits exuded a profound fragility. The peculiar evolutionary histories of oceanic and island seabirds had produced some of nature's most fantastic organisms, but just as certainly, they were organisms that would easily fall victim to the heedless onslaught of human history.

Visitors who had toured the hall when it was completed in 1953 had experienced quite a different image of the Pacific during the previous decade. World War II introduced a Pacific theater to the American consciousness that was sometimes at odds with the literary and artistic representations of the Pacific as paradise. Far from the scene of beauty and tranquility represented in the bird groups, the Pacific was more often associated with war, death, atomic bombs, forced migrations of indigenous islanders, and the terraforming bulldozers of the Fighting Seabees. The actual Pacific had undergone massive ecological changes well before the 1940s, a consequence of the long history of European and Asian colonization of the islands. This process was dramatically accelerated as the Pacific became America's new western frontier and a post-war military buffer zone. A visitor to the Hall of Pacific Birds wrote to Murphy that she was "amazed at the beauty of the settings. We felt as though we had actually visited some of the spots that are in the headlines today, and now we can think of them in terms of their real natural beauty and charm, instead of just devastation and death."[45]

The Hall of Pacific Birds was the showpiece of the Whitney Wing at the AMNH. The wing was funded by the Whitney family and the city of New York, an arrangement that dated back to 1928, when Dr. L. C. Stanford convinced H. P. Whitney to bankroll a new wing for the then overcrowded ornithological collection.[46] The department inherited several massive bird collections from museums around the world, and a great many of its cabinets were devoted to the birds obtained during the famous Whitney South Sea Expedition in the 1920s and 1930s. The new Whitney Wing also boasted a fully equipped laboratory on the eighth floor for the study of bird behavior. The wing opened in mid-1939 with the Hall of Pacific Birds only half complete. Murphy's own history of oceanic exploration was well represented: one group was dedicated to the petrels and albatrosses of the Roaring Forties, and two to the guano birds of Peru. Late in 1942 came the dedication of four new groups that represented landscapes in New Caledonia, Solomon, Fiji, and the Australian Barrier Reef. The war in the Pacific put a stop to the hall's progress as travel to the region became impossible and the staff's time was swallowed by war duties. Cornelius Whitney felt that the cessation of work was "a great disappointment. . . . I felt very strongly that the public interest in the South Pacific Ocean at this time was very keen."[47]

Two years before America had declared war against Japan, Robert A. Falla, then a premier ornithologist and director of the Dominion Museum at Wellington, New Zealand, spent a month at the museum's collections studying its specimens of subantarctic birds. He admired the ocean geography approach of Murphy's *Oceanic Birds* and was attempting to do a similar analysis of New Zealand's population of ocean-dwelling organisms. He and Murphy became immediate friends and collaborators, especially because the western Pacific possessed many of the petrels and albatrosses that had been Murphy's primary research interest since the *Daisy* expedition. Falla suggested two New Zealand locations that the museum would be wise to consider for the hall.[48] Shortly after America dropped two atomic bombs on Japan, the hall committee made plans for several expeditions to find the material for the five remaining spaces. Murphy volunteered to take charge of the New Zealand groups, and in 1947 he set off for the two islands with the twin objectives of securing exhibit material and conducting further research on subantarctic seabirds.

Murphy boarded a train bound for San Francisco, stopping for a few days in Topeka, Kansas, to visit his son and get a haircut. He noted how Topeka was being rapidly built up: "a new and shiny shopping center threatens to make the shop of the nice barber even more of an anachronism." He delivered a paper on the "forests, water-table, vegetation, and animal life in North America as the colonists found it, of what has since taken place, and of the new and necessary trend in which the children will have a part."[49] On his way to San Francisco, he brushed up on New Zealand natural history. He read the story of Pelorus Jack, a white dolphin of Cook Strait that became famous among mariners for its playful antics off the bows of passing ships. In 1904, Murphy took note, the New Zealand government passed an injunction that protected the species because the Maori had constructed a cultural fabric of legends and rituals around the grampus. He also read literature from the Waipoua Preservation Society that was spearheading a movement to make a National Park out of the "virgin" Kauri stands of Waipoua Forest in North Auckland. The original twenty-eight-million-acre stand of native conifers had been whittled down to thirteen at the hands of early settlers and was being supplemented by plantings of North American pines. The entire affair was marked by bitter controversy between preservationist and utilitarian concerns.[50] Soljak's history of New Zealand brought Murphy up to date on contemporary theories of indigenous settlement. He then boarded a steamer bound for New Zealand, but he stopped at Pago Pago to find that "everything was as it should be. . . . The houses looked as airy, and almost as neat as the first explorers found them."[51]

For one of the museum bird groups, Falla and Murphy had chosen animals from the Snares Islands, a series of hard-to-reach promontories off the southern coast of the South Island. The islands were important because of the complete absence of humans. Here they could gather material to construct a representation of an environment that "has never been occupied by human inhabitants . . . and remains in its wholly primitive state."[52] The second site was on the shores of Lake Brunner in Pyramid Valley Swamp, a famous South Island location where preserved moa skeletons had recently been found—a perfect place for executing Murphy's intention "to depict primitive New Zealand bird life of a time antedating European settlement."[53] The objective was to create a diorama that reconstructed a New Zealand scene of some five centuries ago, before the European introduction of invader species.[54] Murphy secured one small moa and an extinct rail-like takahe; when he returned to New Zealand in 1949 he obtained the skeleton of the large moa that was reconstructed in the hall.

The two excursions, and the consequent museum dioramas, describe a moral tale about colonization and conservation. The Snares group was intended to depict a pristine landscape untouched by humans and invading species. Its peculiar island flora and fauna were the results of long evolutionary histories. The message was clear: here is a natural world that still exists, but only because the Snares's inaccessibility prohibited human colonization. In contrast, the Pyramid Valley exhibit depicted a historic landscape. The evolutionary oddity of a completely wingless bird was the product of the moas having lived in an environment free from predatory pressures (Murphy often described the oblivious behavior of penguins, which display almost no fear of humans, in the same manner). The rapid introduction of human beings and their companion organisms sounded the death knell for New Zealand's most peculiar fauna.[55] The conservation story written into the structure of the dioramas was closely paralleled by Murphy's own concerns about conserving New Zealand's landscape. In his expedition report to Albert Parr, he noted that "New Zealand has already gone dangerously far in the exploitation of forests, soil and river erosion, a fact well known to the biologists but realized scarcely at all by the mass of the population."[56] This nascent conservationist ethic became full blown when he returned to New Zealand in 1949 to attend the seventh conference of the Pacific Science Congress.

On his second New Zealand sojourn, Murphy served as a representative of the National Research Council's Pacific Science Board at the first postwar Pacific Science Congress. High on the conference's agenda was the promotion of collaborative research into the ecological effects heaped on the war-torn Pacific landscape, as well as designing a plan to bring the tools of science to bear on questions of conservation and politics, issues that became of utmost importance in the rapidly changing cultural, political, and economic milieu of the postwar Pacific.[57] Robert Falla, presiding over the Auckland conference, asked Murphy to deliver the keynote address. The result was a remarkable synthesis of ideas that Murphy had tackled throughout his life; he wrote an environmental history of the islands of New Zealand, an environmental history that made explicit the moral tale of the New Zealand exhibits in the Hall of Pacific Birds.

Murphy delivered the address, entitled "The Impact of Man upon Nature in New Zealand," to a packed amphitheater of some fourteen hundred scientists and citizens. He introduced New Zealand as a landscape that had undergone massive ecological changes, especially in the hundred years antedating the address. These changes were magnified because of the region's isolation, some one thousand miles from the coast of continental Australia. New Zealand, he noted, consisted of incredible diversity, in both life and climate. North Island possessed environments that he characterized as subtropical, while South Island contained communities that were subantarctic in nature. The islands' varied topography and the winds and currents of the South Pacific Ocean conspired to tie island flora and fauna intimately to the ocean, just as in the case of the Peruvian guano islands.

He called historic New Zealand a "primitive" landscape, by which he meant "the pre-European condition, existing when actual ecological disturbances were begun by Captain Cook." Murphy was not so naive as to believe that these primitive islands were static; he noted the occurrence of climatic cycles, and plant and animal successions. In terms of flora, New Zealand had attained a climax, "a term not to be understood as an ultimate state but rather as a long-enduring state of succession." He also noted the changes in the landscape by the predecessors of the Maori, but "racial or tribal mores, which seem to function so much more purposefully than systems of cause-and-effect perceived by civilized man, held destructive changes to a minimum. Like most primitive folk, the Maori were effective conservationists." Primitive New Zealand was distinctive for its absence of quadruped mammals—again, a function of its isolation. Without rodents or ungulates browsing forest underbrush, all grazing was left to specialized birds, flightless and even wingless birds, which developed to occupy such ecological niches. This environment, a product of New Zealand's oceanic climate and geographic isolation, was also quite fragile. In Murphy's words, it was "preadapted" for organisms like European sheep.[58]

Thence came the opening of Pandora's box when Captain Cook introduced five geese to the islands in 1769. On following visits he sowed cabbage, turnips, and potatoes, and he let loose sheep, pigs, goats, stoats, and fowl to browse the forest underbrush. This biological exchange had many of the same tragic consequences that occur when long-separated people and islands come together for the first time.[59] Murphy chronicled these changes before giving a more optimistic view of the future. More than just conservation and preservation initiatives, he intimated the need for a countrywide change in attitude toward New Zealand. He concluded, "today [the New Zealander] is not only meeting the requirements of the Dominion but is also gallantly and persistently shipping 1,000 tons of food a day to Britain, his spiritual 'home' at the northern antipodes." Murphy thought that if New Zealanders could make New Zealand their spiritual home, then the future of the fragile oceanic island would be a bit brighter.[60] The paper received overwhelming applause.

South Georgia; Peru's guano islands; New Zealand. As Murphy was reaching the twilight of his career, he had already traveled to the ends of the earth and had hewn closely to his 1938 call for "The Need of Insular Exploration."[61] But it was his home of Long Island, above all, that captured his heart. Even as Murphy was taken by the stark wildness of the Snares Islands, so too was he impressed with the natural world in his own backyard.

Murphy's first experiences as a naturalist were in the pine barrens and brackish swamps of his Mount Sinai home on north central Long Island. The stories of young naturalists entering the field with a sense of play and exploration are pervasive in the literature of environmental history. Murphy did likewise, with the one exception that his childhood tramps not only instilled a love of nature, but they also imbued a fondness historically specific to Long Island. Here he honed the journal writing skills that became a lifelong daily practice. Birds, of course, were the object of his truest affection, but nothing seems to have escaped his eye. Before setting off for Brown University, he had amassed a not inconsiderable "annotated list of the species and subspecies of birds known to occur on Long Island, State of New York, and the neighboring waters."[62] After residing in Providence and New Jersey for a time, he and his wife moved back to Stony Brook, where they lived for the rest of their lives. Grace Murphy later wrote of the coastal shingle-sided house with fondness. Speaking for her entire family, she noted that "whatever any of us do we owe at least in part to our home in Crystal Brook. It has seeped through and through the core of each of us."[63] When not on an expedition or working at the American Museum, Robert Murphy could be found exploring his backyard, the Mount Sinai harbor, or on a miniexpedition to various parts of Long Island.

The couple became ardent defenders of the Long Island environment. They played instrumental roles in protecting Fire Island and Gardiner's Island from development, and through the hard work of Grace Murphy's Long Island Conservation Association, the Fire Island seashore saw its way into the state park system. Robert Murphy continuously criticized attempts to dredge coastal regions of the Long Island Sound. In 1966, he vociferously opposed SUNY–Stony Brook's attempt to downzone private land in the village for the erection of multifamily dwellings for temporary student dormitories. He founded a Long Island chapter of the Nature Conservancy. He fought New York's attempt to control the mosquito population through marsh drainage. Finally, he was an energetic litigant in fighting for an injunction against the USDA to prevent the indiscriminate spraying of DDT over Long Island citizens' property.[64] This case was the launching pad for Rachel Carson's *Silent Spring* (1962). Robert Murphy was a prominent figure to the citizens of Long Island. Both his fame as a world traveling naturalist and his efforts to fight for the environmental protection of Long Island made him a source of local pride.

Murphy's final environmental history was that of Long Island. He delivered it in front of the American Philosophical Society in 1962, and it was published in 1964 as *Fish-Shape Paumanok*, also the title of a poem by Murphy's beloved Walt Whitman. He repeated the declensionist narrative of balanced nature to ruined landscape. He could have just as well been talking about South Georgia, one of the guano islands, or New Zealand. He described how the retreat of the Wisconsin ice sheet resulted in Long Island's insular geography. Over time, the island developed a unique flora and fauna with intricate connections to the sea. After exploring Long Island's natural history, he turned to the written record, those archived papers left by the early colonists that told of what they saw, and of how they transformed the landscape. Murphy then told the story of how Dutch colonists, suffering from the "fallacy of the inexhaustible," quickly exploited "the greatest store of natural treasure that has ever fallen into the hands of mankind."[65] Forests were cleared, fauna squandered, game birds indiscriminately killed, and wetlands drained and developed. In the more recent past, he noted, increasing encroachment from New York City's population further threatened the few remaining undeveloped landscapes. In short, Long Island's natural history was one of reckless exploitation, with little regard for *nature's history.* The preface of the published volume spoke of the brilliant success of Murphy's lecture. It noted of Murphy that "in later years, as a naturalist of far-ranging experience, he views these scenes of his youth with deeper understanding, through the eyes of a scientist and student of *human history.*"[66]

The world's various environments—both terrestrial and oceanic—were, according to Murphy, trapped in a destructive human process of discovery, exploitation, and exhaustion. Such an understanding cut to his heart. The destruction of nature also meant the destruction of history, those long geological and evolutionary processes that do the work of creation. Insight into Murphy's heart and mind helps to explain why so many naturalists become environmental activists. Most assume that they do so because of their love and understanding of nature. True enough, but only partly. Naturalists become activists because they view nature as a dynamic culmination of history. And so it was with Murphy's ocean. But other meanings for the ocean wafted through American popular culture during this era. For many, the ocean became the subject of a sublime and sensational spectacle.

CHAPTER THREE

Sensational Management

William Beebe and the Natural History of the Ocean Sublime

Listeners with their radios tuned to an NBC syndicate on September 22, 1932, at 1:30 EDT, heard the daily programming interrupted by the voice of Ford Bond, who was on the deck of the scientific research vessel S.S. *Freedom*, then some seven miles south of Nonsuch, Bermuda. William Beebe, who was the head of the New York Zoological Society's Department of Tropical Research, and Otis Barton, who financed the event, were preparing to enter a steel sphere weighing some five thousand pounds and four feet, nine inches in diameter that was to be lowered in these tropical waters to a record depth of two thousand two hundred feet. Beebe "will literally enter a new world," Bond told the audience, "from which we will bring to you his voice telling of what he sees there." Bond spent twenty minutes setting the scene, emphasizing the authenticity of the event; this was in no way a publicity stunt. "This is a broadcast," he insisted, "of a scientific undertaking. . . . It is not a planned event to make a radio program." Thirty minutes after the program's start, the bathysphere was swung over the ocean, and Bond concluded the first broadcast by saying, "Down . . . down . . . into the depths . . . two men in a hollow steel globe . . . and while you are listening to other programs for the next hour, the Bathysphere will be going down . . . down . . . down . . . down. . . . "[1]

At 3:00 P.M. Bond quickly reintroduced the scene and then handed over the transmission to Beebe, then fifteen hundred feet down, who communicated with his staff on board the *Freedom* through a Bell telephone, and thence through the ether to British and American radios. He was well aware of precisely when the radio transmission began, but instead of acknowledging his new international audience, it was business as usual. His first words: "Four fish going by with orange lights and bluish white color persisting right in path of light. Two eels about two feet long just swam by. Color of water is a bluish-black." For another half an hour, as the bathysphere was lowered another 700 feet to the crushing depth of 2200 feet, Beebe continued to narrate his voyage of discovery by completely ignoring the fact that he was setting a new record; he never noted that his were the first human eyes to peer into the dark abyss of a quarter mile's depth. Instead, he told his audience of the unfathomable brilliance of phosphorescent creatures that poured through the fused quartz windows.[2] Thirty minutes into the broadcast, Beebe made no note of achieving the record of two thousand two hundred feet. Bond concluded the broadcast: "As he has of course been completely absorbed in his work we have had the privilege of participating in real scientific exploration. . . . We have seen this great scientist actually at his work, have heard his reports as they will actually later be correlated and added to the sum total of human knowledge."[3]

This seemingly spontaneous and authentic broadcast of discovery actually masked a carefully choreographed affair. Even before the broadcast, Beebe had made the decision not to address his audience. He wanted to de-emphasize the fact that he was the first to venture into this new abyssal frontier by conveying to his audience that he was not in the business of heroic adventure. His interests were purely scientific. Beebe brought this wish to National Geographic publicity manager, Ford Bond, who drafted a preliminary transcript that was, in turn, edited by Beebe. He wanted to efface his own personality, his own ego, his self-acknowledged status as a hero, from the work of science.

Why did Beebe orchestrate this event? Why was he so concerned with having the dive portrayed as a scientific event? What was the rationale behind his description of the ocean deep as a sublime seascape of wonder and awe? These related questions are all answered by considering the entry of natural history into the realm of popular culture. Andrews went to the ocean as a hunter, Murphy as a conservationist. Beebe entered the ocean as a kind of showman, but he did so with fear and loathing. Despite his purported distaste for the limelight, he desperately needed public attention. Beebe was the director of a fairly substantial research program, and the New York Zoological Society's (NYZS) Department of Tropical Research (DTR) required private funding to keep the operation running. The money came from America's industrial wealth, just like the American Museum of Natural History. But where the American Museum could build temples of natural history in downtown Manhattan, and the New York Zoological Society could put the beneficence of New York's wealth on display at the Bronx Zoo—which they oversaw—the Department of Tropical Research required a presence in the public eye that came through Beebe's nature writing and popular accounts of his exploration work in newspapers, magazines, and books. When the DTR incorporated the bathysphere into its research program, it was almost a certainty that newspapers and magazines would want to sensationalize the dramatic event. And this was the problem.

The dive fit squarely into the genre of "going where no one has gone before" exploration. Beebe received enormous public attention, and while in the limelight, he reported his observations to America. He painted a portrait of an ocean that had only been picked, prodded, and scraped by scientific instruments. Even today the feat sounds extraordinary: a half mile down, he was sitting in a steel ball that was suspended from a ship by a single cable. If this wasn't the stuff of heroism, then what was? But heroism, for Beebe, was a double-edged sword. On the one hand, it made for incredible reading and enormous public attention. But this type of heroism had the tendency to appear a bit foolish—a needless risk, a reckless stunt, clearly not the stuff of serious objective science. And the ideal of objectivity weighed heavily on Beebe's mind. Beebe was faced with an incredible dilemma. How could he publicize the heroic nature of the dives and at the same time maintain his scientific credibility? How could he make himself—his own ego—invisible in order to highlight the ocean itself? The radio broadcast is one example of his larger effort to manage sensationalism. His nature writing, his descriptions of the ocean itself, presented another opportunity.

Those who did not hear the broadcast may have come across Beebe's bathyspheric work in articles written for *Harper's Monthly*, *National Geographic*, the *New York Times*, or in his popular book *Half a Mile Down* (1934). All of his articles, books, and representations of the ocean constitute a literature of discovery that had the net effect of subliming the ocean abyss. Beebe's representations of the ocean fall within a genre central to what students of American nature writing call the pastoral impulse. This kind of nature writing includes such notable naturalist authors as Ralph Waldo Emerson, Henry David Thoreau, and John Muir.[4] These writers transformed the American landscape from a frontier of gloom to one of glory, from a region of fear and loathing to one of spiritual rejuvenation. Essentially, they helped to transform the idea of wilderness into something that was valuable and worth protecting; indeed, some historians agree that the flurry of conservation reform, especially the creation of national parks, at the turn of the century was, at least in part, a function of this new value placed on the American landscape. What these writers did for the American wilderness, Beebe did for the ocean.

Including Beebe within this list of nature writers is problematic for the simple reason that Beebe, like other scientist-adventurers, achieved public attention because he was the epitome of the modern explorer, and explorers' tales are rife with the glorification of human progress, the human penetration of an unknown frontier, the activities of human prowess and heroism, of putting one's physical body in danger for a greater cause. Narratives of heroic exploration and discovery are often the antithesis of humble nature writing.

Moreover, such spectacular events seriously jeopardize the scientific nature of the explorers involved, a dynamic firmly set in place by the Frederick Cook and Robert Peary dispute over the discovery of the North Pole in the early 1910s.[5] Ironically, Beebe's use of the sublime was a consequence of his status as a hero; more precisely, this way of writing about ocean functioned to distance Beebe from the sometimes odious stuntlike nature of heroic adventure. By using the trope of the sublime, Beebe further effaced his own ego, his own existence, to highlight the inconsequential nature of the human condition in the context of a terrifying and vast natural world.[6] But why not just take Beebe's invocation of awe and wonder and an oceanic sublime as a true expression of his experiences? Certainly, it was. The problem is that reports of discovery are rarely innocent acts; they are suffused with culture, values, and a rich texture of social history. To understand how Beebe represented the ocean is to also understand Beebe the explorer, the history and nature of natural history, the growth of popular culture, and an important moment in environmental history when the ocean became a spectacle.

Beebe was from a family of modest wealth whose ancestry hailed back to seventeenth century New England. He was raised in the then rural setting of East Orange, New Jersey, where he began his career—as is so common in the naturalist's tale—observing and cataloging the insects and especially the birds of his local environs. He was an average student who excelled in the natural sciences. After graduating from high school, he attended several courses at Columbia University, where he studied under Henry Fairfield Osborn, Frank Chapman, and Franz Boas, the luminaries of the Columbia-American Museum of Natural History network. Osborn quickly befriended Beebe and secured for him a position as assistant curator of birds at the New York Zoological Society (the Bronx Zoo) in 1899. His primary responsibility was to maintain the aviary. But in 1909, Anthony Kusar, a wealthy society trustee, offered to finance a seventeen-month-long expedition to the Far East to prepare a monograph on the ecology of pheasants, an opportunity that required Beebe to forgo his responsibilities at the aviary. The pheasant expedition was just the first of a life of exploring the biology of the world; from this point on, almost like a bird in seasonal migration, he would split his years between New York and the sites of various research expeditions.

Beebe's environment of choice was the tropics of Central and South America; he thrived in the humid heat of the jungle. Several trips to Venezuela and Mexico hatched in his mind a plan to establish a more formidable scientific presence in the tropics by building a scientific research station. Beebe sought and received the support of Theodore Roosevelt, who, besides possessing presidential fame, was also an amateur naturalist who frequented the New York

City social spaces occupied by Beebe, Murphy, and Andrews. Thus the newly established Department of Tropical Research, including Beebe and a small staff of naturalists and artists, set off in 1916 to establish a semipermanent research station in the tropical jungles of Guyana. Between 1916 and 1926, with a few exceptions, Beebe and his staff summered in the tropical jungles of Kalacoon, and later Kartabo, British Guyana. After locating a suitable stretch of tropical wilderness, the staff set up shop in the former home of Guyana's Protector of Indians. The house was fitted with modern equipment and converted into a wilderness laboratory. The purpose of the DTR, as described in an *Atlantic* article, "A Jungle Laboratory" (1917), was "not to collect primarily, but to photograph, sketch, and watch them day after day, learning of those characters and habits which cannot be transported to a museum. This had not been done before; hence it took on new fascination."[7] The distant laboratories of the DTR became a favorite research site for naturalists from American universities.

Beebe was perpetually concerned with the financing of the operation. Funding generally came from New York philanthropists, the New York Zoological Society, visiting scientists, and Beebe's own pocket.[8] Publicity was the key to all sources. Money came into the DTR only when Beebe presented the impression that it was doing valuable work. In a sense, he was the ad man of natural history; the fate of the DTR and its tropical research stations relied on Beebe's popularity and his ability to market the department's contribution to scientific knowledge. Beebe soon realized that nature writing could perform an invaluable service in publicizing the work for the Department of Tropical Research. His *Atlantic Monthly* editor prodded him in that direction by noting that "there ought to be some way of publishing your essays which should call them to a brand-new audience. The difficulty has been that you are pigeonholed in the public mind with some of your fellow naturalists of the *Auk* variety; whereas you are really nothing but an observer of life and manners among animals and men, who can put poetry into prose."[9] It did not take Beebe long to find his voice, and essays poured from the typewriter of his jungle laboratory desk.

From time to time, he assembled the essays into books that were favorably reviewed. Theodore Roosevelt wrote a glowing review of *Jungle Peace* (1918) in the *New York Times Book Review:* "Nothing of this kind could have been done by the man who is only a good writer, only a trained scientific observer, or only an enterprising and adventurous traveler. Mr. Beebe is not one of these, but all three."[10] By the early 1920s Beebe had earned a reputation as one of America's great naturalist authors, a reputation that opened up new scientific projects as New York's moneyed elite saw fit to provide him with the patronage necessary to expand his research interests. In Beebe's own words, "one millionaire gave me a yacht, another millionaire gave me [another] yacht, and the Governor of

Bermuda gave me an island. I spent ten years under water."[11] Beebe was entering a new phase in his career as he began leaving the jungles of South America behind for new wilderness areas.

In 1923 Harrison Williams, a member of the Zoological Society's Board of Managers, donated his yacht, *Noma,* for a five-week expedition to the Galápagos Islands. Beebe added several members to the staff of the DTR and proceeded to follow in Darwin's footsteps in search of clues to the mystery of evolution. Immediately upon his return, another trustee, industrialist Henry Whiton, offered the steam yacht *Arcturus* for an oceanographic expedition that would investigate the life forms of the Sargasso Sea and the Humboldt Current. Harrison Williams then offered to finance the conversion of the *Arcturus* into an oceanographic research vessel. Beebe had no trouble raising similar funds from others within the New York banking elite. When the expedition was announced in the papers, hundreds of letters from around the country came into Beebe's office requesting positions on the ship. By 1925, Beebe had raised over $250,000 to make the oceanographic venture possible.

THE *ARCTURUS* EXPEDITION

Beebe had already learned the tricks of the trade when it came to managing small tropical research stations. But he had not yet become aware of the importance of managing sensationalism, a skill that would be necessary because his jungle writings had created an audience that ensured a popular following of the *Arcturus* expedition. The expedition was nothing short of a mammoth undertaking. The ship weighed some twenty-four hundred tons, was retrofitted for deep-ocean sampling, and had a scientific staff that numbered fourteen and a total crew of fifty-one. This was Beebe's first foray into marine oceanography, but his objective was very much the same as his jungle work: to describe the life histories of as many species as possible in the wilderness. The Sargasso Sea was thought to be the home of countless fishes and crustaceans, and the Humboldt Current was already famous for its rich and abundant marine life. Beebe called these regions a "wilderness of water."[12]

Throughout the six-month expedition, major metropolitan newspapers across the country followed Beebe and his staff with rapt attention. Much of this coverage was the result of the work of a New York publisher and publicity man, George Palmer Putnam. Putnam was the middleman of the expedition, a service he provided, at least partially, in exchange for permitting his two children to travel along. He received Beebe's correspondences via wireless and telegram and then relayed the stories to newspapers and magazines. Just after the expedition was under way, Henry Fairfield Osborn, director of the American Museum of Natural History, reported to a patron that

> The Arcturus Expedition gives full promise of making history. The breadth of interest is almost mystifying. It owes its existence, in the first place, to the excellent publicity which George Putnam has been largely responsible for, but of course above all to the natural curiosity aroused in all kinds of people by a scientific expedition of this nature. . . . There will be slight question that the expedition will prove to be a very unusual success and of broad meaning publicly and scientifically.[13]

This publicity was important to the patrons who had fronted the money for the expedition. Although Beebe enjoyed writing articles and books for mass appeal, he detested the business of day-to-day publicity.

Putnam was fully aware that Beebe was interested in distributing tasteful, dignified, and nonsensational accounts that highlighted the scientific value of the expedition. But Beebe was not always satisfied with Putnam's handling of these matters. In fact, there was trouble from day one. The day of the launch was well covered by newspapers and magazines that played up the dangerous nature of the expedition. The ship was crowded with friends and family saying their good-byes, reporters taking photographs, crews preparing the ship for departure, and amidst it all Beebe attending to last-minute details. Several reporters pressed him for a comment on the danger of braving "the port of missing ships" (the Sargasso was speculated by some to be the resting place of countless ships mired in the weed). Beebe responded in a manner that would later prove uncharacteristic: "I hope there will be some danger, for we are all looking for thrills. But what the dangers will be, where we shall begin our investigation, or what we shall find, I don't know."[14]

The copywriters for the *New York Times* seemed to anticipate the incipient danger. After the ship had crossed the Panama Canal for Pacific waters, no word was received from the *Arcturus* for eleven days. The *Times*' front-page article was leadingly titled "Beebe Ship Silent 11 Days as Radio Calls to Her in Vain." Robert Cushman Murphy came to the rescue and reported that the erratic weather of the Humboldt Current was probably interfering with the wire communications.[15] Of course, nothing had happened to the *Arcturus*, but the sensational story raised Beebe's ire, and he put it to Putnam for an explanation. Putnam responded in defense, "Very definitely I had nothing whatsoever to do with it. I mean, in no wise was it a 'publicity stunt.'"[16]

The bon voyage event also created another problem for Beebe. In the hustle and bustle, reporters had homed in on four female members of the crew, Isabel Cooper and Helen Tee-Van as scientific artists, Elizabeth Trotter as assistant in fish problems, and Ruth Rose as historian. A picture of the four women, dressed to the nines, was later printed in the Sunday rotogravure of the *Times*. The paper did not make much of the picture, but to Beebe and Madison Grant (then president of the NYZS), the picture sent an impression that the *Arcturus* expedition was more a pleasure cruise than a serious scientific endeavor. For the rest of the expedition, Putnam was hard pressed to prevent similar information from getting to the public. Responding to another Sunday rotogravure spread, Putnam noted that "it is all dignified and absolutely beyond criticism. It is good too from the standpoint of the feminine element, being suppressed. You are quite right there. We will watch that all very carefully."[17]

It may appear that Putnam and Beebe were engaged in a conspiracy to write women out of the expedition, just as nineteenth-century western writers tended to ignore the presence of women on westward expeditions.[18] But Beebe did so less for his own aggrandizement than out of a concern with how the expedition was portrayed to the public, and he had little confidence in the public's ability to recognize women as participants in science, although he had high praise for their work. But he wanted to make sure that publicity was professional and dignified, and this meant that he had to create representations that were dignified in the eyes of patrons and NYZS board members.

The media extravaganza tended to wash over the many problems encountered by expedition members. Beebe had hoped to find an enormous meadow of sargassum weed serving as home to a rich community of ocean fauna. Only sporadic strands of weed were found and the myth died hard, especially to reporters, who pressed Beebe upon his return: "Did you find any signs of ships that have been trapped by the seaweed?" "No, but we saw enough of the Sargasso Sea, after cutting through it three times, to make it certain that there never was any basis for that legend."[19] Beebe had the double misfortune of seeking out the Humboldt Current during an El Niño event. In contrast to Murphy, Beebe did not believe that the absence of the Humboldt was an interesting research problem. He worked quickly to justify the positive scientific findings of the expedition. In a quintessentially Beebe reflection, he noted that "so many things in life come to us obliquely; the road we plan to follow to a certain objective may be the straightest of lines, but an accident may deflect our way into a bypath that proves to be a Road of Destiny."[20]

That road was constructed with the same technologies that one would have found on an older oceanographic expedition, like the *Challenger* expedition of the 1870s.[21] *Arcturus* was equipped with a deep-sea dredge that scratched and retrieved little portions of the ocean floor at a depth of some three and a half miles. Nets were secured to a trawl that collected specimens at various depths. The ship itself was outfitted with a boom walk that extended over the port side from which depth soundings and general observations were conducted. Perhaps the most startling innovation was the hinged pulpit that could lower an observer to water level off the bow. The ship steamed from place to place, occasionally stopping to dredge and trawl for a fixed period. Position, depth, and temperature were noted at these stations; specimens were then brought on board, separated, identified, and preserved as artists and photographers worked quickly to make representations that captured the form and the all too ephemeral color of ocean fauna.

Much of the newspaper coverage of the voyage came from Beebe's reports of strange and exotic fish: "These incredible creatures, painfully secured from their eerie, horizonless world, would be beyond the inventive power of the wildest imagination." Nevertheless, Beebe gave it a shot. He reported finding the living fossil amphioxus; cyclothones with detachable jaws and luminescent teeth; the hatchet fish with a strangely telescoped head; the pharynx fish, which has a distensible stomach; and hundreds of fish with marvelous appendages and bioluminescent organs. These were Beebe's "grotesques, dragons, and gargoyles. Even the briefest acquaintance with these organisms made the fairies, hobgoblins and elves of Dunsany, Barry, Blackwood, and Grimm seem like nature fakery."[22]

Despite the findings, he was quick to call attention to how fragmentary such an exploration must necessarily be. He asked his reader to imagine a race of beings living in the upper atmosphere that had not the physiology for terrestrial life but had an insatiable curiosity for what existed below. On airships they would crudely lower hooks and nets and retrieve only the slimmest representation of terrestrial life.[23] This object-subject reversal is a classic Beebe move; it is reminiscent of his many attempts to see organisms from the organisms' point of view, which always served as a check on human self-importance. "When we find ourselves in an egocentric mood such as this," Beebe wrote about successfully capturing a silver hatchet fish, "we have but to think what comment *Sternoptyx* would make on our own figure were we to drift down past him in the darkness of his deep home."[24] The reversal also reveals Beebe's desire to bodily enter the aquatic realm.

Indeed, *Arcturus* naturalists had such an opportunity with the help of a diving helmet, the use of which Beebe believed was the most important discovery of the expedition. He recalled "trembling with terror, for I had sensed the ghastly isolation" while struggling against a bad swell on the steep slope of Tagus Cove in the Galápagos. Beebe's explanation for writing these "personal digressions" was "to make real and vivid in the mind of the reader, the unearthliness of the depths of the sea."[25] He compared it to interstellar space, then to the moon and Mars. When Beebe pulls out of these soliloquies, he gets to the task of describing the clarity of the water, beautiful lava-sculpted undersea mounts, encounters with tiger sharks, and a host of life forms. He would bring down bags of bait and let tropical fish feed from his hand. Occasionally he made use of dynamite caps to stun specimens long enough for easy retrieval. Beebe enjoyed this more than any other activity. It was his true medium, experiencing nature's wonders in situ.

A final highlight of the expedition was the lucky eruption of two volcanoes on one of the Galápagos Islands; it also provided Beebe with another opportunity to pay homage to the backers of the expedition. Beebe's description of the chance event is saturated with the rhetoric of the sublime. Filled with "wonder and awe," he put the incident into a cosmic context: "I watched an open artery of Mother Earth pouring into the sea-rock liquid as blood. The Galápagos was being born again. . . . The cosmic splendor of the whole thing was overpowering. . . . We had been brought close to the very beginning of things,—and this could not be written or spoken, hardly thought indeed, but merely sensed as one stood apart in a lonely corner of the deck."[26] This description demonstrates a writing style that Beebe often used called biocentrism, which is simply a way of writing about the natural world that places life, instead of humans, at the center of everything. But here he goes further. While describing the Isabela volcanoes, he waxed eloquent about cosmic beginnings, a sense of being overpowered, the failure of language, and human smallness. Beebe also capitalized on the chance event by naming the twin outbreak after the expedition's primary patrons.

In contrast to his jungle laboratory, coverage of the expedition did not make it into many magazines. The entire affair was handled by the daily press, and less than a year after his return, Beebe published a popular book length account, *Arcturus Adventure* (1926). Although it falls short of being a running narrative of exploration, it is a collection of essays that highlights the main events of the trip. Reviews of the volume were mixed. Many gave the typical thanks to Beebe for making science palatable and enjoyable for a lay audience. Of special interest were Beebe's accounts of helmet diving, but it was not his descriptions of the coral-edged Galápagos environment that drew attention. Instead, commentators concentrated on his encounters with sharks. One reviewer noted that the "under water life included sharks, which Mr. Beebe, with a courage no battlefield could surpass, trusted would not attack him."[27] Although such courage received approbation from one point of view, it also had the potential to bite back on account of the sensational nature of such displays. One literary critic who had nothing but praise for Beebe's jungle books leaned toward ambivalence when reading of a new species of deep-sea fish, *Diabolidium arcturi* Beebe. The reader "begins to dread the worst: that his trusted guide has succumbed to the

William Beebe sitting on a coral reef off Nonsuch Island, Bermuda, in the early 1930s. Before the invention of scuba, an air helmet with a rubber hose tethered to a boat was the only way to spend time observing the underwater world. Courtesy of the Wildlife Conservation Society.

enemy's snare at last, and now bids for popularity by purveying sensation."[28] Beebe overcomes the problem, in this reader's eyes, because what is fantastic and sensational is really just a habit of deliberate observation.

Others were less forgiving. "William Beebe is in a little danger of being spoiled," wrote a reviewer for the *Nation*. "He has learned that he is an interesting man who does exciting things; and the poison of self-consciousness threatens to make of him a showman exhibiting himself. . . . Sometimes he writes like a celebrity in a dress suit condescending to a cultured audience which has paid $5 a head to look at him."[29] Professional scientists were even more reserved in their praise. The trouble with Beebe, according to a tropical naturalist, was that he looked at "everything in nature as an 'adventure,'" a trait that he thought "distasteful." He also challenged Beebe's tendency to exaggerate the dangers of exploration. In short, "Too much poetry and too little science, too much adventure and too little calm thinking . . . are the most evident faults of Beebe's writings."[30] Beebe must have taken this sort of criticism to heart. He worked the jungles of British Guyana for another four years, with the exception of a 1927 cruise to Haiti's coral reefs. In 1929 he moved the research station to Nonsuch, Bermuda, the site where his management of sensationalism was tested to an extreme.

ABYSMAL RESEARCH: THE BATHYSPHERE

The *Arcturus* expedition catapulted Beebe into the public eye, but his famous bathysphere dives made him a hero. The bathyspheric work was actually a small part of the Bermuda Oceanographic Expedition's broader research agenda to explore the oceanic life off the coast of Bermuda. Financing his previous oceanic expeditions required massive funding from wealthy patrons, and the nature of such expeditions was always fleeting and piecemeal as oceanographic vessels briefly explored a series of research stations along the ships' itinerary. In 1929 an offer from the Bermudan government made it possible for Beebe to combine his earlier strategy of forming tropical research stations with his newfound fondness for oceanic life. Sir Louis Bois, governor of Bermuda, offered the use of Nonsuch Island off the south coast of the Bermuda mainland for the establishment of an oceanographic and marine biology research facility. Here Beebe would continue to explore oceanic fauna, but the Bermuda station made it possible for an intensive (as opposed to extensive) survey of life within a prescribed area. The offer seems to have been a result of Beebe's March 1929 lecture on undersea life that had piqued the interest of members of the Bermudan parliament and tourists alike.

The establishment and maintenance of the station was dependent on the goodwill of the Bermudan government. "Never," Beebe reported to Madison Grant, "in any country, have we had equal kindness and generosity from the government. All the fishing laws have been abrogated, signal flags have been arranged for communication with the shore, our landing place has been cleared of boulders, etc. etc."[31] Beebe consistently nurtured the station-government relationship by inviting politicians to tour the new facilities. As a consequence, he found easy access to the Bermudan parliament when the station was in need. In April 1929, Beebe convinced the parliament to pay for raising the hull of a sunken fifty-foot tug and moving it to the station's harbor as a wave break.[32] In the 1931 season, the Bermuda governor, Sir Astley Cubbit, and his wife visited the station three times. On one occasion, Beebe let their daughter helmet dive to four fathoms.[33]

Despite the pro bono services of the Bermuda government, managing the oceanographic station was a costly business that required Beebe's constant attention. Seines, deep-sea nets, steel cables, winches, and steel drums constantly broke down or snagged on the ocean bottom, causing Beebe to send a steady stream of equipment orders back to America. In the 1931 season, the all-important steel drum—required for deep-sea trawling—broke under the stress of use. A cable arrived from a New York manufacturer indicating that repairs would take six weeks. Realizing that two months of deep-sea downtime was unacceptable, Beebe steamed to New York, saw to having the drum ready in ten days, and in the process convinced the president of the company to cut the repair bill by two-thirds.[34] The Bermuda station received some funding by the Bronx Zoo, but the lion's share continued to come from patrons, such as Mortimer Schiff and Harrison Williams, who donated $30,000 in 1929. The tropical amenities of Bermuda were also put to good use as a summering spot for the yacht-going wealthy. Beebe reported that he was "particularly fortunate in the number of wealthy men who came to St. Georges in their yachts especially to visit us. Every one of them were enthusiastic, and showed it either in immediate cheques or in a promise of future help."[35] A typical visitor could expect a tour of the facility by Beebe, a trip out on a tug for a day of deep-sea trawling, or even a helmet dive in Nonsuch Harbor.

By May of the first season, Beebe had organized seven departments for scientific research: photography, painting, aquariums, trapping and fishing, technical laboratory, shorefishes, and deep-sea fishes. Beebe was most proud of the activities in this last department. Similar to the *Arcturus*, the DTR's ocean vessel, *Gladisfen,* was outfitted for deep-sea dredging and trawling. Beebe was doing the same kind of research here as he had done in the jungles of British Guyana, where he blocked off a distinct quadrant of jungle space and systematically described all that it contained. A steady stream of scientists and philanthropists visited the research station, and Beebe continued to write of the littoral and deep-sea fauna in both scientific and popular papers. In sum, organizing the station was a sizable task, and that fact did not go unnoticed. A Bermuda copywriter reported that the first impression gained by a layman visitor is surprise at the technical organization. A scientist is frequently imagined to be a somewhat absent-minded person so deficient of ordinary common-sense that he is a prey to the commercially-minded business man. Dr. Beebe is certainly indifferent to commercial or financial success, but as an organizer he can hold his own with a disciplined soldier or a captain of industry.[36]

Since the *Arcturus* expedition, Beebe had received many proposals for designing technologies for deep-sea exploration. But it was not until Otis Barton—engineer, amateur naturalist, and inheritor of a sizable fortune—presented a feasible design with an offer to pay for the sphere's construction that Beebe considered expanding his work in the ocean's deep to include in situ exploration.[37] The bathysphere was cast and fitted with three eight-inch-thick windows of fused quartz, the clearest glass ever produced (as Beebe incessantly told the press). A new tug was chartered and rigged to safely lower and retrieve the six-ton sphere. Beebe designed a meticulous protocol for the DTR staff and tug crew that ensured everyone's safety. Moving picture footage, for instance, shows several of the crew banging away with tools on the steel cable that was bringing in the bathysphere after a dive, a procedure that removed the corrosive seawater from the cable to prevent rusting. By May 1930, the Department of Tropical Research was ready to seamlessly integrate bathyspheric research into their wider program.

A considerable amount of time went into preparations, but the dives themselves only took up a scant part of three weeks in the entire 1930 season. The sphere was lowered fifteen times over the course of seven days. Four descents were test dives in which the sphere was lowered empty; Beebe and Barton were the occupants on the remaining descents but for one in which DTR scientists John Tee-Van and Gloria Hollister were lowered to four hundred nineteen feet. The bathysphere was equipped with a high-energy light, provided by General Electric, and a Bell telephone through which Beebe would relay his observations to Hollister, who took notes from the tug's deck. Indeed, Beebe often received this equipment free of charge so long as the companies had free rein to advertise the usefulness of their products—half a mile under the sea, that is. Four of the eleven manned dives were actually contour dives. Instead of a deep vertical descent, the bathysphere was lowered to between eighty and three hundred feet as the tug slowly moved to give Beebe and Barton a horizontal profile of Bermuda's insular shelf. Beebe consistently emphasized his traditional deep-sea and Bermuda shore research and reported on the bathysphere work with great modesty. Newspapers did not immediately seize on the fantastic nature of the dives until it was announced that Beebe and Barton had reached a depth of 1,426 feet.

Some critics immediately believed they were witnessing a contest for a world record and minimized the descent by citing that miners had been working in the bowels of the earth up to 5,200 feet; "Dr. Beebe must therefore let out more cable to capture the world's depth record."[38] Beebe was quick to disabuse the public of the notion that he was out for a record and began what would be one of many argumentative maneuvers that portrayed the dives as a serious scientific endeavor. Shortly after the first dive to 1,426 feet, Beebe reported that he had seen many fish "attesting to the scientific value of the apparatus," and upon his return to New York, he remarked that "the importance of this deep dive is not the fact that it is the furthest man has ever been under the sea but the great value to science in its study of deep-sea inhabitants." At first, there was little response from fellow naturalists. Henry Fairfield Osborn was easy to convince, and he wrote an approving letter to *Science* attesting to the scientific value of the bathysphere. Beebe was no doubt concerned with winning others. After reading the *National Geographic* account of Beebe's 1930 dives, E. J. Allen wrote a glowing letter of praise. Beebe circulated the letter among his staff with a little note written in the margin: "This scientist ranks close to Osborn and Einstein in reputation."[39] Beebe was equally pleased with an encyclopedia article on deep-sea research written by John Nichols, the curator of recent fishes at the American Museum of Natural History. Nichols claimed that the "bathysphere provides Science with a great metal eye with which it should be possible to see deep sea fishes for the first time in their proper perspective. . . . We may reasonably expect that it will initiate a notable advance in our knowledge of deep sea life by making possible better correlation of what we know of it by other means."[40] This was precisely the kind of legitimization Beebe desired from scientific circles.

Whether he liked it or not, Beebe was becoming a hero of science. A Bermuda reporter interviewed Beebe shortly after the 1930 season and wrote an article entitled "The Modern Marco Polo." The reporter tapped into an interwar sentiment that the age of heroes was at an end, writing, "it is customary among the critics of our time to lament the passing of the age of robust doers of magnificent deeds and the substitution therefore of an age of effete scribblers with no heroic exploits to celebrate. Beebe is a living refutation of that theory."[41] It is unclear what Beebe said to this reporter to make him write that "he dislikes people who lionize him," but more certain is that Beebe was ambivalent if not hostile to his heroic status.

This Bermudan reporter was not alone in lionizing Beebe and in praising his heroic efforts. But this was a heroism of a very new kind. The nineteenth-century hero was a man of great physical strength and mental prowess. The arctic explorer, for instance, had to encounter and overcome a hostile environment through a combination of resolve, intelligence, and physical strength. Beebe was in the same camp as other interwar heroes, like August Piccard and Charles Lindbergh, whose adventures were intimately connected to technology.[42] All Beebe had to do, after all, was sit in a cramped sphere; his greatest discomfort was sitting on a monkey wrench throughout the entire first dive. A science writer for the *New York Times* put his finger squarely on the new dynamic when he claimed that the ascents of Piccard and the descents of Beebe matched the exploits of Columbus, Magellan, and Cook. Their adventures required not only "great physical courage," but they also needed to have a "profound knowledge of physical science involved in what must be called artificial adaption to the environment." Beebe did not possess this knowledge. He relied on the expertise of Barton and other engineers to construct a piece of technology that could safely withstand enormous undersea pressures. Beebe was thought to be a hero, but this was a new kind of heroism by technological fiat.[43]

Beebe became aware of some of the pitfalls and potentials of the bathysphere as an object that mixed science with dramatic spectacle, and he increasingly found new ways of striking a reasonable balance. The bathysphere underwent repairs through the 1931 season, and the unique offer from NBC to broadcast a dive gave him special incentive to use the bathysphere during the 1932 season. Through the medium of radio, Beebe was thus able to bring his research to an entirely new and widespread audience, and his management of sensationalism took on new form.

The broadcast was judged a great media success. An executive from NBC immediately fired off a cable saying that the "program today [was] one of the most thrilling if not most thrilling I have ever heard." The special programs director of NBC noted that "carrying on your work without apparent regard to radio audience proved one of the most graphic broadcasts ever presented." Some critics included the broadcast in lists of the most pivotal radio events of 1932, which included the Lindbergh kidnapping, Japan's General Honjo speaking directly from the battlefield in Manchuria, and Amelia Earhart's greeting to America on her arrival in London after her transatlantic flight. Earhart's husband, George Palmer Putnam, played a large role in publicizing the event. Newspaper writers were often more impressed by the description of the deep sea than the record depth. The *News Chronicle* reported that "New Yorkers, sitting at home with pipe and slippers have been brought into direct touch with the marvels of sea life, half a mile below the surface, by the broadcasting of Dr. William Beebe's record descent in a bathysphere. Amid the roar of Broadway's traffic late yesterday they heard the scientist describe in cool impersonal tones his journey into terrifying depths of the ocean, where eternal darkness is studded by the myriad lights of phosphorescent fishes."

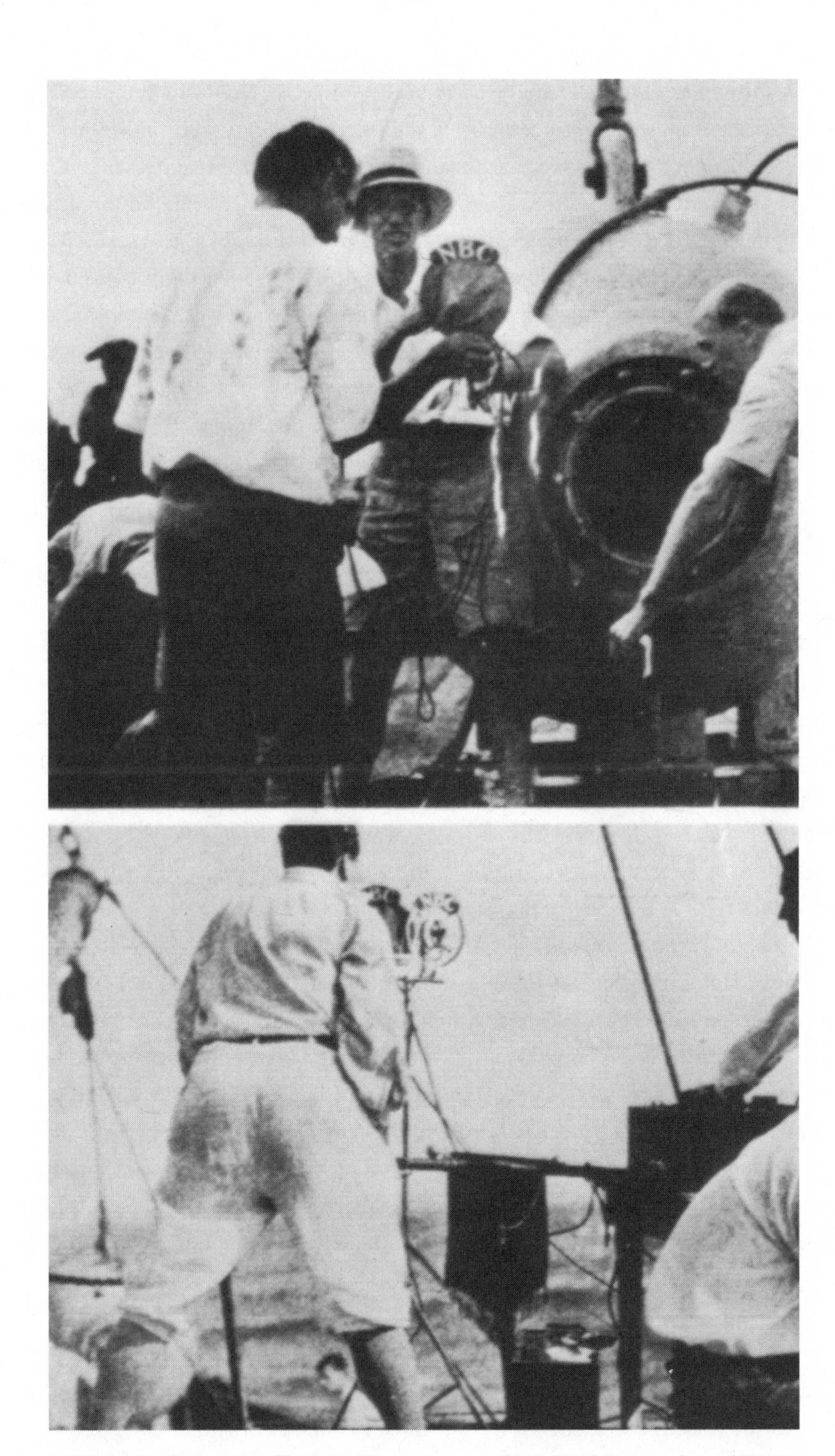

William Beebe and staff of the Department of Tropical Research and NBC making preparations for a live broadcast of a descent in the bathysphere in 1932. Courtesy of the Wildlife Conservation Society.

Despite the overwhelmingly positive coverage, Beebe continued to emphasize science over spectacle. To an interviewer's query about the popular dive, Beebe responded—with a bit more than a hint of hyperbole—that "from above they sent down word that our feat was being broadcast by radio, but my assistant and I were so busy we didn't have time to think of the broadcast until we were told thirty minutes later that it had ended. Sealed up as it were, it seemed utterly impossible for me to think of any one else except my assistant and myself." The interviewer concluded with a note of which Beebe must have approved: "Popular writer that he is, possessing the gift of taking scientific fact and transcribing it into fascinating literature that the layman can understand and enjoy, William Beebe is primarily a scientist."[44] This was exactly the kind of publicity that Beebe was after, and what could have been interpreted as a publicity stunt was portrayed as a serious scientific endeavor.

Beebe also took matters into his own hands by writing up a lengthy account of the dive for the *New York Times Magazine*. The entire goal of the dive, he explained, was to get below the level of humanly visible light where bioluminescence could be directly observed: "I could now prove without doubt whether continued observations from a window such as this would yield valuable scientific observations, or whether the attainment of these depths must be considered in the light of merely a stunt, breaking former records."[45] Beebe voted for the former and went on to describe the strings of siphonophores, coiled pteropods, a school of mychtophids, and other organisms, all of which possessed incredible lighting organs, probably for feeding and mating functions.

Beebe must have considered the whole affair a stunning success. Telegrams from the states congratulated him on the spectacular scientific event. Newspaper coverage was nationwide. Despite the ever-increasing popularity, the sphere sat dormant through the 1933 season, but when the National Geographic Society offered to finance a considerable portion of the DTR's 1934 season, Beebe once again lowered the bathysphere into Bermuda's waters.

The lessons that Beebe learned about publicity and sensationalism during his earlier expeditions were put to good use in managing the publicity for the 1934 bathysphere dives. His increased popularity made careful management absolutely essential. While still in New York late in 1933, Beebe had received word from Gilbert Grosvenor that the National Geographic Society was interested in offering $10,000 to sponsor three or four dives. The event was to be called the National Geographic Society–William Beebe Expedition, and in return for the funds, Beebe would produce two articles with sixteen deep-sea illustrations produced by the artistic staff. Beebe set a number of conditions of his own. First, given the danger involved, Beebe was to be "absolutely in command and having whoever you send as photographer being under my direction."[46] As his prior experience with public photographs bears out, Beebe was really attempting to control the kinds of pictures to be taken and distributed. The second condition was that Beebe would not be asked to dive deeper than he had in 1932—that is, that he not be asked to set a record. Last, the formal name of the venture would be the National Geographic Society–New York Zoological Society Expedition, thus omitting any direct reference to Beebe.

Grosvenor's offer came just in time. The DTR was experiencing financial problems, given the almost ubiquitous impact of the Depression, and Beebe needed extra funding to carry the staff through the 1934 summer. Beebe immediately wrote to Madison Grant, simultaneously asking him for permission and telling him what was about to occur: "This sum will be sufficient to enable me and my staff to carry on our researches for three or four months beyond the bathysphere work, and as I cannot possibly spare a penny more of my own this year, it seems a godsend at this particular time."[47] Not everyone in New York was happy with the setup. Reid Blair, director of the Zoological Park, wrote to Grant that he hoped Beebe "will be successful but I am anxious about this stunt and its effect on the Zoological Society if any accident should occur."[48]

Beebe became increasingly tired with the chore of publicity manager as the successive dives received more and more public scrutiny, so he wrote to an events coordinator at the National Geographic Society and informed him that he was having "information problems." Grosvenor responded to Beebe by offering the services of John Long, the publicity manager for the National Geographic's recent stratosphere project. Long would serve as a conduit and manager of information between Bermuda and the mainland. Freed from this chore, Beebe could devote more time to the management of his own scientific staff. More important, Beebe assigned Long the task of transmitting a particular kind of information. He later wrote that "Mr. Long has been very patient with my desire to be certain that the outside reporters would send unsensational accounts."[49] Long's management was even more propitious in that the society's press releases were distributed to five hundred papers worldwide. Such an advertising network would bring massive publicity to the event, and it would be publicity more or less under Beebe's direct control.

With the NGS dives scheduled for August, Beebe and his staff spent the summer working diligently in Bermuda while two high-altitude balloonists, Captain Stevens and Major Kepner, were preparing to best Jacques Piccard's stratosphere record. The National Geographic Society also sponsored this expedition. Long managed the publicity and suggested that Beebe radio good luck to the duo in the interest of creating an aura of comradeship for the balloon and bathysphere expeditions. In a humble prelude to a more famous conversation between explorers in Sea Lab and Space Lab in the early 1970s, Beebe penciled a note on the telegram: "Heartiest good wishes to Stevens and Kepner for splendid observations on stratosphere flight and a happy landing."[50]

Long arrived in Bermuda early in August, and plans were made to dive sometime in the second week of the month. The bathysphere, marred from its transport from the Chicago Exposition, where it sat in juxtaposition to Piccard's balloon, received a fresh coat of aquamarine paint. The sea calmed enough for an easy descent on August 11; Beebe and Barton were lowered to a record depth of 2,510 feet. Over the receiver, Beebe could hear the *Ready*'s steam whistle and shouts of congratulations when the sphere had passed its previous record depth. But Beebe put a halt to the dive just short of the half-mile mark, which was the expected goal. Beebe told reporters after the dive that "the only reason he did not go on to the half-mile goal was that he had been more interested in making observations at various levels than in merely setting a record."[51] In the excitement of abounding life, Beebe had "completely forgotten the idea of a half-mile record."[52] He did not forget four days later when, with only a dozen turns' worth of cable left on the winch, the sphere was lowered to the deepest mark it would ever reach, 3,028 feet. Aside from one shallow dive for John Tee-Van, the bathysphere research of the 1934 season came to a quick end, and it also marked the conclusion of in situ deep-sea research at the DTR. Beebe would never again find himself bolted into the narrow confines of the bathysphere.

Long had kept the presses fed with developments up to that point. After the dives, he prepared a long release that reported some of Beebe's more spectacular observations. The release emphasized the record depth, but Long was quick to postscript the news with a by now standard disclaimer: "While one of the objects of the expedition was to go down a half-mile, it was not solely record depth that Dr. Beebe sought. Before, between, and since the two record dives, he and his aides have made many dives to observe sea-life at various depths."[53] Long was attempting to clearly define the objective of the dive in contrast to media coverage that often explicitly stated that *the* goal of the exploration was to dive to record depths.[54] Beebe was happy with Long's work, but he ranked the entire publicity effort by the National Geographic Society as only "75%."

But for a few quibbles, the 1934 season was another splendid success. Publicity was widespread and more or less dignified, and the National Geographic Society received considerable praise for sponsoring the expedition. Reid Blair, the director of the New York Zoological Park, who was earlier worried about Beebe's "stunt," congratulated Beebe on the dives and went on to say that he too wished that "we might benefit by the success of your spectacular dives. We have not had a single new member since June 1. I am not envious of the Geographic's success but wish we might have benefited to some extent."[55] In all fairness, Beebe had never failed to mention his affiliation with the New York Zoological Society. Indeed, the society had at least a little cachet as a scientific institution that Beebe always used to increase his own scientific capital.

LEGITIMIZING THE BATHYSPHERE AND USES OF THE SUBLIME

Beebe was constantly trying to find ways to improve the scientific value of bathyspheric dives. At one point he sat down to write a note, likely to himself, on "Bathysphere observations especially wanted." First on his list of general observations was to note the color of every animal, and he castigates himself with a side note, "not 'cheek light pale' or 'body covered with multitude of small lights.'" He wanted to observe the precise mechanisms of fish locomotion and schooling behavior. He also created a list of specific questions. Do cyclothones "move in unison like miniature surface schools?" "What does Chaulidus's first dorsal ray do?" "What is the source of a shrimp's luminescence?" "Do they twist their eyes back?" He finished on a note that by this time had become the quintessential Beebe style: "Try to concentrate in off moments on the small stuff—number of copepod schools etc."[56]

Other than refining his own bathyspheric observations, Beebe made inquiries to physical scientists for advice on possible experiments so as to put the bathysphere to good use. E. O. Hulbert of the Naval Research Laboratory, for instance, responded by giving a long list of experiments that would investigate brightness, spectrum analysis, polarization, and cosmic rays. Furthermore, he suggested that Beebe contact Robert Millikan, of cosmic ray fame, for further possibilities. Beebe wanted to mirror the stratosphere flight's use of a spectrograph, and probably would have done so if it weren't for the bathysphere's physical limitations.[57]

One way of proving the scientific value of the bathysphere, or of disproving the stunt, was to stay at the bottom of each dive for an extended period of time. Beebe did much to publicize the extended periods of observation that these dives required. These trips were more than a quick jaunt for a record. Beebe was hoping "to stay in the depths of the ocean for five or six hours at a time, photographing and studying the inhabitants."[58] Then he came up with a new way to use the technology. The bathysphere, towed by a ship, could swing low over Bermuda's underwater plateau, and thus Beebe could construct a highly accurate topographical map of the seafloor. Unfortunately, the winch was not able to raise and lower the sphere fast enough for the new method to be of much practical value.

Despite Beebe's constant tinkering with the technology, the bathysphere could not rise above one insurmountable problem: crucial to the practice of the natural historian was the collection of actual specimens. For taxonomists like Beebe, the pickled specimen in hand was necessary to do the work of science. In the bathysphere, he was little more than a voyeur. Beebe even attached baited hooks to the outside of the sphere, with no success. When he tried to name a new species of fish that he had identified through the bathysphere's quartz window, a host of naturalists cried foul.

Several scientists were less than enthusiastic with the bathyspheric work. They were even more ambivalent about the popular account of these dives, *Half a Mile Down* (1934), in which Beebe attempted to name new species and genera from his observations through the bathysphere window. Hugh Darby found fault with Beebe's attempt to write a popular scientific account: "The weaknesses of his book are due largely to the fundamental incompatibility of their demands."[59] Carl Hubbs responded similarly but took Beebe even more to task in reporting and naming seemingly impossible fish. "I am forced to suggest that what the author saw might have been a phosphorescent coelenterate whose lights were beautified by halation in passing through a misty film breathed onto the quartz window by Mr. Beebe's eagerly appressed face."[60] The value of the bathysphere as a valid scientific technology was under fire, and even the director of the Zoological Park could not "see much prospect of his obtaining any real scientific data from such dives." The book was less than a scientific text than one that should "take its place in the annals of daring exploration."[61] Beebe acquiesced to the pressure. Shortly after the expedition members returned to New York in 1934, he told reporters that "the dives had no scientific value" and that he would devote his time hereafter "to diving in places where the sea and land merged and where specimens for observation and capture were to be had."[62]

Beebe's quest for legitimizing the bathysphere moved beyond technological tinkering and a tight control over publicity. He also could do it through his descriptions of the oceanic abyss. The clearest manifestation of this influence was Beebe's invocation of the sublime. The sublime is an emotional aesthetic that is felt when one experiences a spectacle of unfathomable and ineffable awe—the view from atop a mountain or from beneath a thundercloud. Immense scale is the key to this emotion. Immanuel Kant referred to the failure of the imagination to present objects to the understanding. This is a function of experiencing a spectacle that overburdens the senses. Terror, desolation, and isolation are also associated with experiencing the sublime; this is one of the keys to distinguishing the sublime from the beautiful. Experiencing the sublime is also coupled with a certain failure of language. A visitor to the 1907 electrical lighting of Niagara Falls responded, "words fail to describe the magnificence of the spectacle." Having no reference that would compare with the view of Niagara, another visitor declared that the spectacle was "unearthly." All these associations with the sublime serve to shrink down the human ego. Sublime feelings highlight the insignificance, the minuteness of the human condition in the presence of nature's most vast and awe-inspiring spectacles.[63]

Beebe was not the first writer to represent the ocean as a sublime spectacle. Romantic poet John Keats often wrote of the terrible and awe-inspiring nature of the ocean beating against the shore. Even Sigmund Freud attempted to explain the expression of religiosity by calling attention to an "oceanic feeling"—"a sensation of eternity, a feeling as of something limitless, unbounded—as it were, 'oceanic' . . . it is the feeling of an indissoluble bond, of being one with the external world as a whole."[64] J. W. Van Dervoort's *Water World* (1883) represents the movement of this poetic and psychological imagery into the vernacular. Van Dervoort set out to write a book on the little-known and seldom visited ocean realm for the American public. In something of a religious treatise that glistens with the phraseology of the argument for design, he set out to remedy the paucity of popular oceanic writing. "This is most singular," he noted, "when it is considered that this majestic ocean, whose mighty heart throbs in sympathy with the pulse of God, covers more than three-fourths of the whole globe, habitable and uninhabitable."[65] Filled with a spirit of reverie, awe, and humility, *Water World* was nevertheless very much a nineteenth-century text. It remained for Beebe to articulate an ocean sublime that was literally three-dimensional.

Feelings of the sublime are usually linked with a visual experience where a natural phenomenon overwhelms one's imagination. But Beebe also invoked the discourse of the sublime while discussing the most elemental danger of deep-sea diving: pressure. The crucial element of the immensity of scale is provided not only by the visual, but also through the awareness of deep-water pressures—the immensity of the ocean pressing in on body or bathysphere. On the fourth dive of the 1930 season, Beebe was at fourteen hundred feet when the tiniest semitransparent jellyfish floated past the view of his quartz window. Otis Barton casually told him that every square inch of glass on the window was receiving six hundred fifty pounds of pressure. The spectacle of this fragile creature living under such great pressure dizzied Beebe. He wrote, "I had to call upon all my imagination to realize that instant, unthinkably instant death would result from the least fracture of glass or collapse of metal. There was no possible chance of being drowned, for the first few drops would have shot through flesh and bone like steel bullets."[66] The immense power of abyssal pressure was amply demonstrated during the first test dive of the 1932 season when the bathysphere, *sans* explorers, came up full of pressurized water. The thirty-pound wing nut to the door was loosened and then the door shot clear across the deck as if from a cannon. "All my life," Beebe noted, "I had heard of the terrific pressure at great depths and had seen bottles and cans come up crushed, but never until now had I had first-hand visual proof of this phenomenon."[67] In almost every popular account, Beebe persistently comes back to the feelings of awe and terror that he experienced while helplessly trapped within his steel ball.

But more than evoking just the pressure of abyssal depths, Beebe also invoked the sublime while experiencing the strange colors of this new world. He was observing the bioluminescence of several fish during the 1930 season when he experienced "a tremendous feeling of awe at the astounding glory of glowing lights and iridescence, and never-ending marvel of a living fish" that can withstand deep-sea pressures.[68] Beebe was also moved by the brilliant blue ambient light that filtered through the window. At six hundred feet on his first dive, he "brought all my logic to bear . . . and tried to think sanely of comparative color, and I failed utterly. . . . I think we both experienced a wholly new kind of mental reception of color impression. I felt I was dealing with something too different to be classified in usual terms."[69] As the ball descended and blue gave way to black, he was awed by the "the terrible slowness of the deepening shade."[70]

Searching for similes and earthly reference was quite difficult. Beebe compared the experience with an astronaut attempting to describe a new planet for the first time. When pressed for a comment on his special interest in the dives, Beebe responded that "the depths of the sea below the level of light was so bewildering and amazing that all I can say is that I shall watch for anything that comes within view with the same ignorance and excitement as a rocketeer would have for the Martian faunas."[71] Beebe consistently compared the experience and his observations in the bathysphere to space exploration, still only a possibility for fiction writers and a few engineers in the 1930s. He simply was nonplussed with the task of describing what by all appearances was a new world. The problem was not lost on Rudyard Kipling, one of Beebe's favorite authors, who reported that he had "been reading that amazing Fourth Dimensional book of yours that you so kindly sent me—Half Mile Down—and I find I haven't any scale for its measurement. It looks like the first opening up of a new world . . . like Columbus throwing fits on the discovery of the Western Tropics."[72]

Beebe's most powerful invocations of the sublime occurred on a general level when he described how he felt a quarter or a half mile below sea level. Far from a tale of heroic conquest, Beebe continuously called his readers' attention to his own insignificance within the ocean deep. In characterizing the first season of diving, he said that he "shall never experience such a feeling of complete

Argyropelecus and *Sternoptyx* shown as the frontispiece of the *Arcturus Adventure* (1926). The artistic staff of the Department of Tropical Research became adept at displaying the wonders of the deep sea for presentations, books, and magazines. Courtesy of the Wildlife Conservation Society.

isolation from the surface of the planet Earth as when, a few months ago, I dangled in a hollow pea on a swaying cobweb a quarter of a mile below the deck of a ship rolling in mid-ocean."[73] When the telephone went out on a dive at two hundred fifty feet, Beebe responded to being cut off from all communication: "We had become veritable plankton. I visualized us as hanging in mid-water. . . . The silence was oppressive and ominous, and our whispers to each other did nothing to alleviate it. The dark blue outside became cold and inimical."[74] The day after his very first dive, he sat down to reflect and crystallize what he observed "through inadequate eyes and interpret with a mind wholly unequal to the task. To the ever-recurring question, 'How did it feel?,' etc., I can only quote the words of Herbert Spencer, I felt like 'an infinitesimal atom floating in illimitable space.'"[75] Beebe reaches the height of self-effacement when he compares himself to floating peas and infinitesimal atoms—a strange image given that Beebe was a heroic explorer of a new world. But this latter fact was exactly what Beebe wanted to distance himself from.

In 1936 Beebe wrote a short reflection on the deep-sea dives. His intention, he remarked, was to study the nature of light—or, rather, the lack of light—in the ocean deep. His early visits brought him to levels where a small amount of sunlight created an ominous and brilliant blue. Beebe, again reflecting on and rationalizing the dives, indicated that going deeper was necessary to break into the realm of total blackness:

> In 1932 and 1934 I gave a half-mile as the answer to innumerable questions, but I cared nothing for any man-made measurements of depth—my only goal was to get beyond solar light. . . . And instead of a swift descent and return and only a silly record to brag of, we were staying minutes and hours in the sphere, comfortably looking out and seeing creatures which had evolved in the blackness of a blue midnight for the past one hundred million years or so, ever since the second day of creation when the oceans were born.[76]

There is no direct evidence that Beebe invoked the sublime to mitigate the unpalatably sensational nature of his descents. But in this slender piece of evidence, we see how Beebe's concern that the bathyspheric dives be portrayed as a thoroughly scientific endeavor—in no way a spectacular stunt—gives way to sublime thoughts on the creation of all life. Beebe's use of the sublime, just like his choreographies with Putnam and Long, his technological tinkering with the bathysphere, even his hobnobbing with diplomats and industrialists, played an important role in the drama of managing sensationalism. As a consequence, his ocean was dark and gloomy, mysterious and confusing, awe-inspiring and humbling.

OTIS BARTON, THE BATHYSPHERE, AND THE NEW YORK WORLD'S FAIR

Beebe would later trivialize his entire oceanic program, especially when compared to his more thorough jungle studies. It's almost as if he was trying to disassociate himself from the entire experience, most importantly the record-breaking dives in the bathysphere. His involvement with the bathysphere did not end in 1934, however. New opportunities and problems emerged that once again called for Beebe's management of sensationalism.

Beebe was in no way alone in representing the ocean; Americans could increasingly catch a glimpse of the sea either at aquariums or movie theaters. Before Beebe's work on the *Arcturus*, there were a scant seven aquariums in the United States. Twenty new facilities were constructed between 1925 and 1950, and a remarkable twenty-seven in the following thirteen years.[77] One of the most famous and popular public aquariums during the 1930s was in Coney Island and operated by the New York Zoological Society. Its director, Charles Townsend, a colleague of Beebe's, expedited the explosive growth of urban aquariums by publishing a guide for the construction, equipment, and management of public aquariums. "No other form of public museum is of greater interest to the people," he noted. "The plentifully stored aquarium is an ever-changing exhibition of beautiful and useful living things not seen easily in their natural habitat. . . . It has great educational value, stimulating constant inquiry respecting our heritage of the waters, which a wasteful civilization must take still greater pains to conserve."[78] Another successful aquarium was Marineland, an ocean-themed aquarium constructed in 1938 near Saint Augustine, Florida. Marineland simultaneously served as a public window into the ocean, provided a research lab for scientists, and functioned as an aquatic movie set for motion picture producers.[79] The ocean was also becoming an increasingly important geography in motion pictures like *20,000 Leagues under the Sea* (1916) and *Marine Circus* (1939). These movies opened up an underwater vista that had been completely unknown to American audiences. They did the work of demystifying the oceanic realm, and at the same time they turned the ocean deep into a spectacle of adventure, wonder, and awe. Movies and aquariums played a role in advertising the ocean as a place of beauty, and Beebe thus played a part in this wider move to transform the ocean into a popular culture spectacle, but he did so on his own terms.

When Otis Barton decided to fund the construction of the bathysphere in 1929, he had every intention of taking moving pictures for a major undersea adventure movie.[80] The 1920s were a heyday for the hunting adventure film, and Barton simply wanted to use midocean depths as a new setting for this popular genre. Beebe often had to sacrifice time at the bathysphere window so that Barton could shoot film. Unfortunately, the technology for undersea photography, much less moving film, was still in its infancy. The lighting was poor, and the footage produced nothing notable. Barton moved his filming operations to Panama, where he was able to hire actors to perform some mildly sensational scenes in shallow waters. The Educational Films Corporation of America—specialists in educational, scenic, and scientific subjects—produced Barton's film, which came out in 1939. The movie did contain some shots of the bathysphere from the deck of the *Ready*, but otherwise it took place on the sandy beaches of Panama.

The producers sent a copy to Beebe late in 1938; it was then titled *Bathysphere.* Beebe's only response was to encourage the producers to follow through with their plan to rename the movie to *Titans of the Deep* and to omit all shots of the DTR staff, leaving only the Panama footage.[81] When the film was released in 1939, word around town seemed to imply that *Titans of the Deep* was one of Beebe's projects. He fired off a letter to *Science* distancing himself and his staff from the entire affair.[82] *Titans*, a movie that highlighted staged sharks preying on swimming women, had very little scientific value, according to reviewers. Beebe had every right, as Barton seems to have admitted, to disassociate himself and the staff of the DTR with the film. So even after Beebe had banished the bathysphere from his research program, he continued to preserve its image and those of his research team and the New York Zoological Society as dignified, nonsensational, and, as always, thoroughly scientific. The timing of the release coincided with the New York World's Fair, a venue that presented Beebe with the opportunity to exhibit his science to America's public.

When not employed in the business of exploring Bermuda waters, the bathysphere inhabited public spaces as a testament to the spirit of exploration. Between 1934 and 1936, it stood, as if suspended by the *Ready*, in the American Museum of Natural History's Hall of Oceanic Life. The bathysphere took its place underneath the gondola of August Piccard's stratospheric flight at Chicago's Century of Progress Exposition in 1933 and thus symbolized a new phase of exploration. Humans, long restricted to the plane of the earth's horizon, were now exploring the depths of the ocean and the dizzying heights of the upper atmosphere. In 1939 the bathysphere was proudly displayed at the New York World's Fair. Beebe's involvement with exhibit design demonstrates his desire to stir the imagination with natural and fantastic wonders while presenting the dignified nature of his natural history, at the same time reinforcing his recurring message that humans were only a small and overly egoistical part of the natural world. He also wanted to simulate the bathysphere dives in such a way that the audience itself would become participants, not mere observers, in exploration.

The role of science and technology in fostering economic and democratic progress was a moral at the very center of these fairs.[83] The New York Zoological Society's exhibits powerfully demonstrated the wonderful power of ocean creatures. For example, an electric eel was housed in an aquarium at the society's designated building, and to open the fair, the society harnessed the current from the eel to generate a humble current that ignited the magnesium flares along the Great White Way.[84] Other exhibits by the society contained a series of dioramas that juxtaposed New York's past and present natural history. The former contained many of the creatures Beebe had retrieved from the Hudson Gorge, "animals stranger than the wildest conceptions of our Hollywood dream manufacturers." The diorama also displayed how deep-sea fishes were caught in silk nets and how the bathysphere was lowered into the ocean depths. Close to the diorama were preserved specimens of deep-sea fauna. The bathysphere was located in the bathyspherium, a domed and darkened room, "much as it must have appeared a half mile down to the fishes of the deep," models of which were illuminated by an ultraviolet light.[85]

Beebe reluctantly entered the world of publicity, entertainment, and high adventure. These were activities that had a consistent potential to call his scientific status into question, but they were also the lifeblood of the Department of Tropical Research. As a conductor of a vast enterprise that required careful and meticulous management, Beebe was in the business of courting and pleasing an eclectic audience that ranged from New York Zoological Society naturalists, the wider scientific community, his primary patrons, and a public audience that seemed to watch his every move. At the heart of the dynamic was the persistent problem of balancing science with sensationalism, heroism with humility, egocentrism with biocentrism, and adventure with serious study. He achieved such ends by carefully controlling his and his staff's public image, as well as by composing nature writings that mitigated the stigma of fruitless adventures and reckless record seeking.

Despite all of his passion for science and nature, it is difficult to label Beebe a conservationist or an environmentalist. He was very seldom active in conservation reform. Beebe articulated his relationship to conservationism in a 1948 annual meeting for the New York Zoological Society. There were three kinds of people, he noted: the great mass who are selfish and thoughtless about the natural world; a small minority of conservationists; and then a thin sliver of naturalists like he and his team: "In our work with Birds, Beasts, and Butterflies, all we ask is to be completely divorced from Humans, although we are grateful to the Conservationists for trying to keep Wild Creatures alive for a few more years, so we can study them."[86] But Beebe's writings did deliver a critique of modern civilization. As opposed to many of the participants in the back-to-nature movement, Beebe did not go to the wilderness for retreat; he went there as a student of nature. He watched with some despair the heedless rush of modern development that threatened those wilderness spaces. Behind this development was a human self-confidence—Beebe called it humanity's "misplaced self-importance"—in the ability to control and manage nature for utilitarian ends. This attitude fostered a biocentrism in Beebe's writings that did more than offer a nonutilitarian view of nature. He invested in nature an inherent value that was in no way contingent on humans. Viewing nature, for Beebe and his readers, had an inherent aesthetic and spiritual value. Nowhere is this more evident than in the oceanic frontier. A reviewer of one of Beebe's Bermuda books noted that for most Americans, fish are simply the potential ingredients for a "nourishing bouillabaisse," but in Beebe's hands, these routine observations are "the elements of tremendous dramas, high adventures."[87]

These adventures were not solely the province of men of science; they were activities that could be enjoyed by all. Time and again, Beebe would give a recipe for this kind of appreciation: "Get a helmet and make all the shallows of the world your own. Start an exploration which has no superior in jungle or mountain . . . provide yourself with tales of sights and adventure which no listener will believe—until then he too has gone and seen, and in turn has become an active member of the 'Society of Wonderers under-sea.'"[88] Beebe introduced to the American public this new world, one that he had visited himself: an ocean full of spectacular wonders that inspired feelings of awe and wonder, a world that Americans would flock to shortly after the war with mask and snorkel. As yet the thought of polluting or destroying these natural wonders occupied the minds of only a few select people, like Murphy and Charles Townsend. When all is said and done, Beebe made the ocean something worth protecting—and protection it would need. World War II and the anxieties of the cold war were about to reshape the way Americans interacted with the ocean. Some of these activities had a benign effect on the nature of the oceans; many, however, exacted an unimaginable toll.

Rachel Carson's *The Sea Around Us*

The Construction of Oceancentrism

At 5:00 A.M. on October 22, 1946, a watchman making his last round at Knutsen Shipyards moved in to investigate a disturbance in the shallow waters of Huntington Bay on the north shore of Long Island, New York. He realized that a fin whale had stranded itself in the bay, and he immediately called the coast guard. As morning broke and news traveled around the community, the whale attracted hundreds of local spectators. Schools let the children out, and in a short time, the beach was crowded with five thousand onlookers. Teachers roamed about, prodding the children to take advantage of the event by noting the whale's shape and behavior. Youngsters in small skiffs circled the whale for closer examination. The coast guard's first attempt to return the creature back to the ocean failed, and by the time the tide came back into the bay, two coast guard cutters and an army tug could do nothing more than tow a lifeless carcass out to sea.[1]

What is interesting about the event is what did not happen. No one tried to haul the whale onto the beach and charge admission for people wanting a closer look; and no manufacturing representatives made a bid on the depository of oil. There were no acts of cruelty, as there were in 1939 when a whale showed up in Long Island Sound near Great Neck, only to be plugged by fifteen rounds from the guns of the Nassau County marine police, or as there were in 1944 when harpoon-brandishing police and citizens chased down a whale lost in Flushing Creek. Such events elicited comment from Harold Anthony, an American Museum of Natural History naturalist, who couldn't understand the "all-too common destructive attitude toward any cetacean coming into local waters."[2] In contrast, the Huntington Bay fin whale became an educational opportunity for an intrigued audience. The coast guard's activities were equally interesting. Before 1936 coast guard cutters sometimes received orders to scare pods of whales away from fragile fisheries with oars and firearms; on one occasion, a coast guard ship assisted a recreational fisherman in corralling a whale for a venture in hunting fun. After 1936 the coast guard became the chief enforcer of American violators of the International Whaling Treaty. But the Huntington fin whale seems to be one of the first instances of the coast guard's attempt to *rescue* a beached whale.[3]

The event sounds familiar to the modern ear. There is no greater evidence of the intimate bond between humans and whales—between land and ocean—than the heroic and often legion efforts of humans coming to the rescue of wayward cetaceans. Such activities were fairly new to the postwar period, and they speak to a new relationship between Americans and the ocean. The meanings and uses of the ocean that were established in the early twentieth century by the likes of Andrews, Murphy, and Beebe accelerated in scale and scope during the latter half of the century. Americans continued to treat the ocean as a geography of sport and adventure, as a trove of resources for exploitation and perhaps conservation, and as an object of popular spectacle. But the ocean began to take on new meanings and new uses. Once again, ocean explorers found ways of reinforcing and criticizing that new relationship.

Like a great flood tide, the ocean rushed into the bays and coves of postwar American culture. It became the setting for popular war literature, such as James Jones's *From Here to Eternity* (1951), Herman Wouk's *Caine Mutiny* (1941), Nicholas Monsarrat's *Cruel Sea* (1951), and James Michener's twin epics *Tales of the South Sea* (1947) and *Return to Paradise* (1951). The ocean was both nemesis and seat of spiritual renewal in man-in-the-raft narratives like Robert Trumbull's *The Raft* (1942), Thor Heyerdahl's *Kon-Tiki* (1950), and Ernest Hemingway's *The Old Man and the Sea* (1951). Reflective perspectives of the ocean, like Anne Lindbergh's *Gift from the Sea* (1955), pointed to the nurturing power of the sea. Science of the sea, too, was represented by Rachel Carson's *The Sea Around Us* (1951), Eugenie Clark's *Lady with a Spear* (1953), John Steinbeck's *Log from the Sea of Cortez* (1951), and Jacques Cousteau's *The Silent World* (1953). Many of these works made it onto the silver screen as dramas and documentaries. They were joined by movies of wartime heroics like *The Fighting Sea-Bees* (1944), starring John Wayne, and the resurrection of the 1930s effort to film Jules Verne's *20,000 Leagues under the Sea* (1954). A number of sea-themed TV programs traveled through the airwaves in the 1950s: *Sea Wolf, Voyage to the Bottom of the Sea,* and later, *Flipper.* Public marine aquariums sprung up around the country in unprecedented numbers. Skin diving and scuba diving became popular recreational activities. Hawaii enjoyed a booming tourist economy. Pacific cuisine could be sampled in one of Trader Vic's campy franchises. RCA issued a new recording of Debussy's *La Mer* in 1951 with new liner notes written by Rachel Carson. Fashion designers began marketing hats with ocean themes and Pacific-style lava-lava wraps. And in the early 1960s, the Hawaiian tunes of Don Ho became a national phenomenon. How does one explain this flowering of ocean culture?

In 1947, at the leading edge of this profusion of oceanic pop culture, the septuagenarian fishery biologist Robert Ervin Coker reflected on his life, dedicated as it was to exploring the secrets of the sea. Amid the glowing embers of postwar cataclysm, he looked at both the ocean and his country and noted that "a more widespread 'sea-consciousness' must prevail in the future."[4] What Coker called a "sea-consciousness" can more aptly be termed a sense of oceancentrism—an understanding that the oceans, not the land, dominate the earth. Clearly, the fact had been noted many times before Coker's statement, but it was not until the postwar period that the common phrase, "the earth is composed of 70 percent ocean," became a matter of vernacular understanding. It is a powerful ideology that lends itself to philosophical musings; just as Copernicus, Darwin, and Freud caused humans to reexamine the nature of humanity, so too did oceancentrism cause humans to think of their earth as an ocean planet with several modest continental islands. To capture the cause of oceancentrism is a difficult task. However, we can explore one of the earliest, clearest, and most popular statements on oceancentrism, Rachel Carson's best-selling book, *The Sea Around Us* (1951). Her life, work, and love for the ocean highlight the myriad meanings that the ocean absorbed in mid-twentieth-century America.

Carson is better known for her popular critique of pesticide use, *Silent Spring* (1962), the work that helped to launch the modern American environmental movement. The epochal nature of this book has overshadowed Carson's popular work as an oceanic nature writer, the crowning achievement of which was *The Sea Around Us,* which stood on the *New York Times* best-seller's list for eighty-six weeks. The power of this elegant work brought Carson to hitherto unknown heights of popularity, and contemporaries and historians alike justifiably view the text as a stepping-stone to *Silent Spring.* But *The Sea Around Us* has a distinct history that merits serious examination. Its chief contribution lay in transmitting an oceancentric depiction of the earth in which humans were insignificant participants in a drama dominated by the role of the ocean.

Where did this idea come from? And why did the message resonate so strongly with a postwar audience? The oceancentric argument of *The Sea Around Us* brought together several distinct streams of Carson's experience with the ocean—experiences that were, on their own, not unique to Carson. Much like the teachers in Huntington Bay, prodding their students to explore the mysteries of the fin whale, Carson was enamored with natural history and nature writing, which vested the ocean with a sense of awe and wonder. Like many Americans, she grew up on the oceanic representations of explorers like Andrews, Murphy, and Beebe. Her exposure to this popular and scientific literature provided a kind of exploration of the ocean through imagination and literature. Her interest, however, was not restricted to the ocean; her chief literary and philosophical influences were a suite of nature writers who sought to minimize human self-importance through biocentric portrayals of nature. The search for an explanation for Carson's oceancentrism does not necessarily entail a journey through geographical space, but is rather an inner journey of a reader, editor, and researcher who filled notebooks with notations and citations that were stuffed in filing boxes with meticulously culled references. The influence of nature writing framed Carson's truest ambition as a writer that, up until 1951 when she quit her government position, weathered the centrifugal force of writing for the machinery of federal conservation.

Like the crew members of the coast guard who attempted to save the fin whale as the tide flowed out of the Huntington Bay, Carson was a government official, a scientific editor for the U.S. Bureau of Fisheries. Carson was privy to the forefront of scientific research during a time when the science of the sea—oceanography—enjoyed explosive growth. She would never be able to boast of an active scientific research program, and she only rarely participated in the modest research activities of so many government biologists at the Bureau of Fisheries, which merged with the Biological Survey in 1940 to become the U.S. Fish and Wildlife Service (USFWS). Instead, she put a public face on the work of the bureau; she was an editor of the bureau's publications and a booster of the bureau's scientific work in playing a positive role in the conservation of marine resources. Although Carson much improved the writing of the bureau's publications, she thought of this work as more of an economic necessity than a heartfelt desire. Nevertheless, it was partly because of this editorial position that she became a master synthesizer who developed the rare skill of writing stories that were at the same time about nature and the work of scientists who explored the secrets of nature.

Finally, like the Huntington Bay children standing along the shoreline trying to catch a glimpse of the fin whale, Carson moved through natural spaces as a consummate saunterer, in the Thoureauvian sense of the word. The sentiment was perfectly expressed by Carson's undergraduate biology teacher, Mary Scott Skinker, who was at the Marine Biological Laboratory with Carson in 1929. The two would spend their free time on short excursions along the beach at Woods Hole. Skinker recalled that there was a mystical quality about these walks: "Rachel would wander off by herself, silently watching the ocean, utterly captivated by the sounds, smells, and rhythm of the ocean as well as by the variety of the marine life all around her."[5] Though Carson also cast the gaze of an inquisitive naturalist, Skinker's comments capture the mode of Carson's personal exploration of the ocean, a kind of inner exploration into the heart and the mind that mingled natural curiosity with an ineffable spirituality. Carson, like so many other nature writers who framed this point of reference, called this a "sense of wonder." The final effect of *The Sea Around Us* was to draw from that new scientific understanding of the ocean, combine it with an older and venerable tradition of oceanic natural history and nature writing, and pepper both with a biocentric sentiment that had long been used by terrestrial-bound nature writers. Carson was forging new ground, a new genre. The book was part science of the sea, part literature of the sea, and a full treatment of the ocean as an object of mysterious wonder that dominates the earth.

At its heart, the writing and popularity of *The Sea Around Us* has a relatively simple explanation. While working on early drafts of the book, she wrote to William Beebe and told him of her "belief that we will become even more dependent upon the ocean as we destroy the land."[6] The statement reveals Carson's belief in a process that was unfolding before her eyes. The resources of terra firma were showing signs of increasing stress and degradation, despite the hard work of conservationists and preservationists that had advocated a new orientation toward nature for some fifty-odd years. Americans, as well as many others, would have to turn to the ocean to sustain life. Militarily, economically, and culturally, Americans were poised to demand a great deal of the oceans that surrounded the North American continent. Carson knew this all too well, and she had mixed feelings about how the process would unfold.

A strange paradox lay at the heart of Carson's first contact with the oceanic realm. Although her oeuvre can be characterized as nature writing, science writing, or a combination of the two, she had early on inhabited institutional spaces that emphasized laboratory science—places that rarely cultivated nature literature. Born and raised in a suburb of Pittsburgh, her first glimpse of the ocean was at the Marine Biological Laboratory (MBL) at Woods Hole, Massachusetts, in 1929, a site that had been and continued to be a home to modern experimental biology.[7] Carson's summer project seemed to fit the bill fairly well. Her lab time was spent comparing the terminal nerves of various reptiles—hardly experimental, but clearly a morphological project that required attentive lab work. When not dissecting reptiles, Carson spent her time walking along the shore examining tide pools. She also spent much time in MBL's considerable library. Here she began culling facts about the ocean that formed the foundation for *The Sea Around Us*. So despite the laboratory focus of MBL life, Carson found time to do the two activities that would dominate her life: to observe nature generally, and to research it in the scientific literature.

Portrait of Rachel Carson while working at the Fish and Wildlife Service. Courtesy of the U.S. Fish and Wildlife Service.

The paradox deepens when we consider that shortly after graduating from the Pennsylvania College for Women, Carson enrolled at Johns Hopkins University, another hotbed of laboratory work, especially for genetics.[8] Rheinart Cowles, then head of Johns Hopkins's new Chesapeake Biological Laboratory, urged Carson to examine the urinary system of fish for her master's thesis. After a detailed analysis of the morphology of catfish embryos, Carson produced a thesis that, in retrospect, sounds a bit peculiar: "The Development of the Pronephros during the Embryonic and Early Larval Life of the Catfish." It is difficult to reconcile Carson's later writings with her lab-dominated experiences at Johns Hopkins. She simply thought that laboratory science was but one of many ways of knowing nature. She would later say that her "real preoccupation is not with 'pure' or abstract science. . . . I am the sort who wants above all to get out and enjoy the beauty and wonder of the natural world, and who resorts only secondarily to the laboratory and library for explanations."[9] Whatever her thoughts on lab science when she finished her master's degree in

1932, her career in academic biology came to a sudden end, mainly because of her taxing family responsibilities, which were exacerbated by the Depression. Carson heeded the advice of Skinker and in 1935 took the civil employee exam to secure work in the U.S. Bureau of Fisheries.

THE NATURE-WRITING TRADITION

Elmer Higgins, then director of the bureau, hired Carson as a feature writer of scientific subjects and gave her the task of writing short radio scripts on marine life for a radio series called *Romance under the Waters.* By 1936 she was writing short introductions for bureau brochures. One piece of writing, however, stood out from the rest of her typical work. The article, "World of Waters," turned out to be Carson's first piece of oceanic nature writing. Higgins told Carson that the article was not appropriate for a bureau publication and suggested that she submit it to *Atlantic Monthly,* the journal of record for nature writing.[10]

Carson submitted a revised manuscript to Edward Weeks, editor and successor to Ellery Sedgwick, the *Atlantic* chief editor who was so fond of Beebe's natural history writing. What Sedgwick saw in Beebe, Weeks had found in Carson. The published essay, retitled "Undersea," contained some quintessential Beebe moves. The second paragraph told her readers that "to sense this world of waters known to the creatures of the sea we must shed our human perceptions of length and breadth and time and place, and enter vicariously into a universe of all pervading water." Carson asks us to envision a traveler who begins on a sandy shore and moves slowly to the ocean abyss, observing life along the way. Beebe had asked his readers to do much the same in his descriptions of undersea life.[11] Carson describes plankton as "boundless pastures. . . . Drifting endlessly, midway between the sea and the air above and the depths of the abyss below, these strange creatures and the marine inflorescence that sustain them are called 'plankton'—the wanderers." Again, she might have had a copy of Beebe's *Arcturus Adventure* in front of her at the time.[12] Finally, without directly invoking Beebe, Carson portrays the ocean abyss with all of his signal cues on cold, blackness, and geologic time: "In these silent deeps a glacial cold prevails, a bleak iciness which never varies, summer or winter, years melting into centuries, and centuries of geologic time. There, too, darkness reigns—the blackness of primeval night in which the ocean came into being."[13] Carson would not meet Beebe until 1938, but given the relationship that flourished in the 1940s, and Carson's later statements that she had read Beebe's work, *Half Mile Down,* four times, it is safe to assume that Carson's earliest oceanic nature writing was deeply influenced by Beebe's writings.

"Undersea" was not, however, merely derivative of Beebe's literature. The article evinces Carson's personal exploration of the sea. Her descriptions of the rhythmic tide's effect on shore life certainly came from personal observation of tide pools and sandy beaches. References to spine-studded urchins likely came from her experiences at Woods Hole. The descriptions of clams, lobsters, haddock, cod, flounder, and halibut stemmed from her work with the Bureau of Fisheries, for Carson noted that "from these and shallower waters man, the predator, exacts a yearly tribute of nearly thirty billion pounds of fish."[14] She does not end, however, on such a utilitarian note. The final message was decidedly cyclical, even ecological. She brings all the elements of the essay together to describe the "cosmic background" in which the "seas continue their mighty and incomprehensible work."[15] The sea draws in chemical elements from land and air; with the sun's energy, these substances are consumed by planktonic animals that in turn are fed upon by shoals of fish that die and are redissolved into simple elements to start the entire process anew.

"Undersea" caught the eyes of Quincy Howe, editor at Simon and Schuster, and Henrik van Loon. Van Loon had been at the forefront of carving out a new literary and cultural space that some historians call middlebrow culture, an effort by America's intellectual elite to draw on the new technologies of mass culture to enlighten the American masses. Throughout the 1920s and 1930s, van Loon had written a number of popular books on the history of western civilization and geography. These were texts that had fully embraced an outline form that owed much of its popularity to H. G. Wells's *Outline of History* and became a key genre within middlebrow culture. In van Loon's own words, this type of literature "carried this necessary information to this new public and in such a form and shape that it could understand this new knowledge and enjoy it."[16] Van Loon became one of Carson's early mentors in the publishing world, and time and again, he gave her advice on how to write for a popular audience. With van Loon's and Howe's encouragement, Carson began outlining an intimate "portrait of the sea" that became her first book, *Under the Sea-Wind* (1941).

In her correspondences with van Loon, Carson made clear how the book would need to hold closely to the primary dictum of nature writing: the inseparability of content and form. In order to convey an ecological understanding of nature's interconnectedness, the book had to take the form of a narrative: "The fish and other sea creatures must be the central characters and their world must be portrayed as it looks and feels to them—and the narrator must not come into the story. . . . It seems to me that the principal thing the book must accomplish is the creation of undersea atmosphere, and this [narrative style] seems the best and generally most agreeable way to accomplish that end."[17] We see here a biocentric sentiment, a desire to tell a story through fish eyes to minimize human egocentrism. This form of nature writing is evident in those writers that Carson admired the most.

Carson's most important early literary influence during this period was the nature writing of Richard Jeffries, whom she referred to as a "literary grandfather." Jeffries was an essayist of the English landscape throughout the latter half of the nineteenth century. His later writings reveal not only a deep sympathy for the natural world, but also highlight the insignificance of human beings.[18] Carson was an admirer of Jeffries's biocentrism and singled out his essay, "Hours of Spring" as particularly important. Here Jeffries relates an instance of waking up in the morning and becoming, as Carson paraphrased it, "aware that the buds are opening on schedule and the birds arriving and he wonders how these things can be happening without him who had always so carefully observed them."[19] Carson held Jeffries in such high esteem that she appropriated a line from his "Summer Pageant" to title her first book.

Henry Tomlinson, a contemporary of Carson's, was an English essayist and book writer on a broad range of subjects including travel, history, criticism, antiwar rhetoric, and fiction. Not the standard nature writer, Tomlinson often turned to nature in an attempt to discover an underlying truth that was somehow out of the grasp of humans. He even compared his literature to the transcendental writings of Henry David Thoreau.[20] When Carson caught wind of Sputnik's successful orbital flight, her thoughts turned to Tomlinson, who "suggests over and over—there must be a Truth that has eluded man's understanding. If only he could grasp it—and its intimations are everywhere about—surely all this could be avoided! It seems to me that what Tomlinson—and I—and others have to say should have been said a generation ago. Is it too late?"[21] If humans could conceive of themselves as but a small part of nature, Carson seems to be saying, then there would be no need for war, violence, and international posturing.

Carson was also fond of Henry Betson, editor of *The Living Age* and chronicler of Maine's landscape. Much like Tomlinson, Betson attempted to find a mysterious force within nature, especially along the craggy shorelines of Maine, that seemed to emanate from the ecological unity of life. For instance, he expressed a deep sense of humility and reverence in *The Outermost House* (1924), one of Carson's favorite natural histories. "We need another and wiser and perhaps a more mystical concept of animals," he wrote. "We patronize them for their incompleteness, for their tragic fate of having taken form so far below ourselves. And therein we err, and greatly err. For the animal shall not be measured by man."[22] While still in the process of writing *Under the Sea-Wind,* Carson made a pilgrimage to the site where Betson had built the Cape Cod house that became the setting for *Outermost House.*

Perhaps the most important influence on Carson's early natural history was the English naturalist Henry Williamson, who had written several narratives from various animals' points of view. Williamson achieved such an effect by anthropomorphizing animals to make them the central characters of narratives that could be easily read and understood by a popular audience.[23] The narrative flow of *Under the Sea-Wind* is accomplished with precisely the same maneuver. Carson fleshes out the life histories of migrating shorebirds through the story of Blackfoot and Silverbar; those of mackerel through the eyes of Scomber; and the adventurous migrations of eels from inland pond to Sargasso Sea through Anguilla the eel.

If Carson drew from Williamson, Betson, Tomlinson, and Jeffries for her narrative approach and biocentric perspective, she came to know the actual life histories of these organisms through her work with the Bureau of Fisheries. For instance, in July 1938 Carson visited the Fisheries Station in Beaufort, North Carolina. When not examining the fishery facility, Carson spent her time examining Beaufort's marsh pools, ponds, and shoreline. She paid special attention to the flow and ebb of the local tides and their effect on seashore life. The notes of her observations became the basis for the first chapter of *Under the Sea-Wind.* She told Edward Weeks at *Atlantic Monthly* that "'Flood Tide' describes the events that take place during the rising of the night tide on a small island on the North Carolina coast. The island actually exists as I have described it . . . and some of the happenings I record took place before my eyes."[24]

Of course, the latter two sections of the book, largely underwater descriptions, could hardly have stemmed from Carson's personal observation; rather, they were taken from life histories pieced together by marine biologists. The Bureau of Fisheries had long been charged with the task of informing America's fishing fleet of scientific knowledge that would lead to the efficient conservation of natural resources. Government biologists investigated the feeding, mating, and migration habits of commercial fishes in order to prevent excessive depletion. Other than the study of fishes, government biologists were also students of the sea itself, examining currents, tides, upwellings, and temperature disparities. In *Under the Sea-Wind,* the adventures of Scomber the mackerel drew heavily from her contact with government biologists, especially those working in Chesapeake Bay.

Carson's readings of nature writing and her work as a government editor saturate the entire text, but it is most evident in her narration of a fishing boat's attempt to capture a school of mackerel with a seine-purse net. Carson deftly describes the schooling habits of mackerel and then moves on to a dramatic narrative in which the mackerels escape the fishermen's grasp by sounding through the slowly closing hole at the bottom of the encircling net. Before the escape, Carson delves into the mind of one of the fishermen on deck, or at least into a consciousness that she wished had occupied the fisherman's imagination. The man had been working on the boat for two years,

> not long enough to forget, if he ever would, the wonder, the unslakable curiosity he had brought to his job—curiosity about what lay under the surface. . . . It seemed to him incongruous that a creature that had made a go of life in the sea, that had run the gauntlet of all the relentless enemies that he knew roved through that dimness his eyes could not penetrate, should at last come to death on the deck of a mackerel seiner, slimy with fish and gurry and slippery with scales. But after all, he was a fisherman and seldom had time to think such thoughts.[25]

If Carson did not intend to write an autobiographical vignette, then she at least put some of her ideas into this fisherman's mind. These are Carson's incongruous thoughts as she tried to reconcile her own position within the conservationist machinery of resource utilization—a position that she was not entirely satisfied with—and the wonder, fascination, and respect that she had for oceanic life. They are also the sentiments that Carson found in her reading of Betson, Tomlinson, and Jeffries.

Although *Under the Sea-Wind* was not a commercial success, the book did receive some critical acclaim by reviewers. An unsigned *New York Times* reviewer noted that humans rarely take the time to examine the ocean: "he knows little even of the sea-wind's birds, much less of the ocean's creatures. It takes a naturalist [like Carson] to look at what is there." William Beebe thought that Carson was at her best when describing complete life histories: "There, her attention is concentrated upon a single individual organism, about which environment, experiences and enemies, are made to revolve, and on which they focus." George Miksch Sutton, an ornithologist at the University of Oklahoma, noted that Carson may be an expert ichthyologist, "but she is primarily a lover of the sea" writing "from a wholly new viewpoint, the viewpoint of a creature that experiences daily the 'eternal rhythms of the seas.'"[26] Henry Betson's review praised Carson for showing how "the sun is always more than a gigantic mass of ions, it is a splendor and a mystery, a force and a divinity, it is life and the symbol of life."[27] These reviews highlight how *Under the Sea-Wind* was being read as a natural history of oceanic life. This was not a book about oceanography, or marine science, or fishery biology. Although Carson may have been using her fishery knowledge of eel and mackerel migrations and predator-prey interactions, the focus was not the science, but rather the ocean itself.

The failure of *Under the Sea-Wind* to capture popular acclaim has been explained by historians, and Carson herself, as an unfortunate consequence of timing. "Poised for the popular reception that she had every right to expect after such glowing evaluations," writes Carson's most recent biographer, "Carson's hopes were dashed by world events [the bombing of Pearl Harbor], which ultimately deprived her of commercial success."[28] Although there is some credence in this interpretation, it assumes that all of Carson's work would inevitably meet with popular praise. It also does not explain the fact that Carson was unable to place any of the book material in popular magazines; even *Atlantic Monthly* turned down a manuscript. *Under the Sea-Wind* was, after all, a work of nature writing, a genre that held uncertain promise in the publishing world, especially for newcomers like Carson. Carson's eventual fame was the result of a book of a different genre, one that combined nature writing with science writing. It also required a new context, a new war, and a new crisis for humanity.

SCIENCE WRITING

After she had finished her master's degree, Carson wanted to continue on to her Ph.D. at Johns Hopkins and then go on to a career as a biology teacher. Economic necessity caused her to leave the university and begin work for the Bureau of Fisheries. It was for the same reasons that she became a feature writer of fishery biology and conservation for the *Baltimore Sun* and, more sporadically, national magazines. These experiences were absolutely crucial in the development of Carson as a writer, for the trade of science journalism was much different from that of nature writers. Although the profession of science journalism evolved within networks outside of government science, Carson's peculiar take on the genre was firmly rooted in her experiences with federal bureaucracy and scientific efficiency.

The U.S. Bureau of Fisheries was commissioned in 1872 as an agency to explore American waters for harvestable fish populations. The late nineteenth and early twentieth centuries witnessed a boom in fisheries that quickly depleted coastal populations of commercial fishes, and the Bureau of Fisheries turned to the problem of conserving oceanic resources so that America's fishing fleets could continue to enjoy a sustainable resource. The bureau either hired or contracted zoologists who went into the field to observe the life histories of fish populations as well as the operations of fisheries. Bureau scientists wrote up reports on their research that were distributed to fishermen and government representatives charged with writing legislation that ensured the sustainability of fish resources.[29] Key to this work was the bureau's research ship *Albatross,* which roamed the waters that surrounded American territories.

When Roy Chapman Andrews steamed to the Orient in 1910, the *Albatross*'s scientific staff was exploring waters for commercial fishes near the Philippines, which had been recently acquired in the Spanish-American War.

Carson was not one of the many federal biologists whose stock in trade was the exploration, conservation, and culturing of fisheries. She was the editor at the Division of Scientific Inquiry, and she helped other biologists prepare and publish their reports. One of her earliest contacts was assistant bureau chief Robert Nesbit, who was working on Chesapeake fishes. Carson helped by analyzing statistical data and writing up reports and brochures for the public. She also had the opportunity to consult with experts in fish biology and visited many of the bureau's field stations.[30] We know little about this period of Carson's life, but we can say with some degree of certainty that she was learning how to write about science for the public. Carson steadily climbed the bureaucratic ladder and in 1949 became the chief editor of all Fish and Wildlife Service publications (in 1939 the Bureau of Fisheries joined the Bureau of Biological Survey to become the U.S. Fish and Wildlife Service). She and her staff arranged tables, edited text, and illustrated and organized reports and brochures that were sent to the Government Printing Office.

Carson's own writing for the Fish and Wildlife Service was very limited, but when Albert Day replaced Ira Gabrielson as director of USFWS in 1946, Carson thought it an auspicious time to propose a series on the national wildlife refuge system, eventually titled "Conservation in Action." Representative of the series was an issue devoted to the general topic of "Guarding Our Wildlife Resources." Carson spent about two years visiting national refuges and compiling facts on the history of America's dwindling wildlife populations. Through brief encapsulations of the historical problems with migratory birds, waterfowl, game mammals, and fishes, Carson came to the general statement that we need to preserve "our wildlife in a modern world, in which the advance of civilization is too often a destructive one."[31] But "Guarding Our Wildlife Resources" is really about the work of government biologists who had identified and attempted to remedy the problem. She addressed the work of bird banders, fish trackers, habitat specialists, and mammal ecologists who were working hard to strike a balance between preserving endangered species and ensuring economic progress, mostly through the establishment of wildlife refuges. The publication was liberally illustrated with pictures of animals in their natural settings, but many are the pictures of government biologists in the field: planting round-stem bulrushes, capturing and weighing waterfowl, placing bands on migratory birds, planting salt licks for elk populations in Jackson Hole, transporting deer and sheep onto ranges with plentiful vegetation for browsing, and sorting through marine specimens seined out of a harbor. Although elegantly written, this publication can hardly be called a specimen of nature writing. It was an exemplary piece of *science writing* that demonstrates Carson's ten years of interacting with USFWS biologists. Although her government position provided Carson with limited opportunities for composing original pieces, her extraprofessional articles in newspapers and magazines provided another venue for developing her skills as a science writer.

There was no dearth of newspaper and magazine articles discussing scientific subjects throughout the late nineteenth and early twentieth centuries. But it was not until after World War I that scientific journalism became a distinctive field, with its own set of codes and standards. This new journalism took shape when it became increasingly obvious that scientists were playing a large role in creating military technology and fostering economic development. An early manifestation of scientists' desire to explain the findings of science was the creation of the Science Service in 1921. With financial support of newspaper magnate Edwin W. Scripps and prominent zoologist William Ritter, the Science Service became an arm of several scientific societies and began distributing articles to more than a hundred newspapers in the 1920s.[32]

Carson was a journalist with a scientific background, but she was not a credentialed scientist with an active research program. Her editorial duties at the bureau put her in an ideal position to write articles for newspapers. Beginning in 1936 Carson began a correspondence with Mark Watson, editor at the *Baltimore Sun,* that resulted in a series of newspaper pieces on conservation, mostly about Chesapeake fishes. Similar to her later "Conservation in Action" series, Carson was publicizing the ways in which government scientists were helping fishermen to mine local waters. Her first article, "Science Keeps Watch over the Sea," demarcated science as the solution to both fishery problems and the wider problem of a mechanistic world gone awry:

> Does this mechanistic trend of civilization threaten the great natural resources of the sea even as it has reduced or destroyed much of the more familiar wild life of the continent? . . . Perhaps because the fisheries belong to a strange element of which we know comparatively little, they are often regarded as being exempt from the forces that have rolled other living species into the abyss of extinction. The fertility of the sea, the mother of life, is pictured as inexhaustible. Fishery biology, born of the present generation, has made marked progress toward its goal of removing the uncertainties that have beclouded the field of production in the fisheries throughout their history, and is providing a basis for the scientific management of the industry. . . . From scientific deduction to practical application is a difficult step. It must be taken with the aid of the industry and with the cooperation of

fishermen often set in the ways of forefathers to whom the comings and goings of fishes were acts of Providence, uncontrollable and unpredictable.... The fishery biologist is interested in seeing the industry placed on a basis of sound and intelligent management comparable to the principles of animal husbandry practiced by every progressive rancher.... It insures against hasty and unnecessary legislative shackles by the level of production compatible with the preservation of the species.[33]

The article reveals much about Carson's frame of mind. First, the same mechanical threat that was reducing the land to waste also promised, according to Carson, to deliver similar damage to the sea. Second, the work of the fishery biologist was comparable to the "progressive rancher" who set the business of animal husbandry on a firm scientific foundation. But what distinguishes this kind of writing from *Under the Sea-Wind* was that the only glimpse of nature in this article was a challenge to the myth of the ocean's inexhaustible resources, a point that Murphy and a few others had been making for twenty years. Although Watson made some significant editorial changes, by and large, the article remained a booster for government biology. Like so much science writing in the twentieth century, this was a statement on how science benefited society through efficient scientific management, a mark of Progressive conservation.

Shortly after, Carson submitted an article entitled "Sentiment Plays No Part in Save-the-Shad Movement: Federal Men Called on to Try to Solve Biological Mystery."[34] Thirty-five years earlier, the participants in the "nature faker" controversy lambasted natural histories that embraced sentimentalism at the cost of scientific accuracy.[35] *Sentimental* remained a pejorative term throughout the early twentieth century, and Carson was thus careful to describe how the effort to preserve shad broods from fishermen was not an emotional crusade but rather a rational and scientific move to conserve a crucial resource. Throughout the 1940s, Carson continued to write on similar marine subjects, including whaling, terrapin and trout farming, and oyster harvesting. Each article informed a public audience of the scientific measures taken to conserve marine resources and thus helped to dispel the myth of the sea's inexhaustible resources. Carson also wrote newspaper copy about terrestrial conservation issues, but the message remained very much the same.

Some of these pieces of science writing do contain moments characteristic of the nature writing of "Undersea." Tellingly, an article on migrating eels that Carson had written for the *Baltimore Sun* was a condensed version of the eel saga that became a third of *Under the Sea-Wind*.[36] In this lengthy Sunday edition article, Carson addressed the mysterious migration of eels from midocean to inland rivers and lakes, but instead of the eel's-eye view of *Under the Sea-Wind*, we get something of a report on what scientists have observed in the eel's life history. In the end, the article was a report of scientific findings on the natural history of the eel. Holding true to form, Carson concludes the article with a statement on the economic importance of eel fisheries in the Chesapeake Bay.

Carson was being pulled in two, sometimes irreconcilable, directions: nature writing, her love and passion, and science writing, her bread and butter. Was there room for synthesis? Quietly and humbly, Carson began etching out a new genre of literature that combined the two. For instance, reports of wartime aeronautic heroics presented an opportunity to write a piece on the extraordinary flying abilities of the chimney swift. The longer draft of the *Coronet* publication presented an account of the swift's mastery of the air. Except for spending the night in the shelter of chimneys and hollow trees, Carson reported that swifts spend their life feeding, drinking, and mating on the wing. The piece then reports the pains that naturalists have gone through to observe these elusive creatures. For instance, George Sutton spent many nights anchored to the top of a chimney and observed wing movements as swifts dropped into the shelter. Carson then reported the work of swift banders who have "now provided the solution of a major mystery of bird migration." Another article that contained this mixture of nature writing and science writing told of the bat's use of echolocation. Given the development of radar and sonar technologies during the war, Carson entitled the article "The Bat Knew It First."[37] In short, Carson was beginning to etch out a new kind of article that combined natural history writing and science writing. At the same time, they contained a sense of awe, respect, and wonder—so characteristic of *Under the Sea-Wind*—and reports of scientific research as were found in many of her newspaper articles and Fish and Wildlife Service publications.

Carson became increasingly dissatisfied with her USFWS job throughout the 1940s. Apparently she found editorial work tedious, and Carson wanted to concentrate on her own writing, or at least hold a job that would allow her to work on her own nature writing. She was unhappy with the conservation work of government biology and the type of literature she was producing for government bureaucracy. Carson therefore asked William Beebe for a writing job at the New York Zoological Society shortly after the end of the war. "Frankly," she wrote, "I don't want my own thinking in regard to 'living natural history' to become set in the molds which hard necessity sometimes impresses upon Government conservationists."[38] In this telling remark, it appears that Carson wanted to leave the world of ocean resources behind in order to write natural histories with a deeper respect for nature. Beebe took the request seriously and talked the matter over with Henry Fairfield Osborn, although no job materialized. At the same time, Carson contacted the editor of *Reader's Digest* requesting a science writing job. She also sent out feelers to the Audubon Society.[39] None of these inquiries bore fruit, and she resigned herself to stay on as editor at the Fish and Wildlife Service while writing in her spare moments. In early 1948 Carson conceived of a new project, another book-length treatment of the sea, though this was to be a peculiar picture of the ocean that combined the philosophy of nature writing with the journalism of science writing.

NATURE WRITING MEETS SCIENCE WRITING

The book was tentatively titled *Return to the Sea,* and Carson wrote late into the night for several years to construct an ocean that combined her personal exploration, the government work of bureaucratic marine conservation, the sciences of fishery biology and oceanography, and the self-effacing spirit of nature writing. Two general overviews of the ocean had then recently been published. Carson told Beebe that she wanted her treatment to lie "somewhere between the books by R. E. Coker and Ferdinand Lane—rather nearer the latter, yet I hope to give it a somewhat deeper significance, while still writing for the nontechnical reader."[40] The remark is revealing. Ferdinand Lane was fast becoming a generalist nature writer with popular books on trees, insects, rivers, and mountains. His *Mysterious Sea* (1947), filled with quotes from Shakespeare, Milton, Matthew Arnold, and Coleridge, was a well-written natural history of the ocean that had much in common with Carson's text. Lane provided a lucid and comprehensive account of the ocean's geologic, physical, and biological history. R. E. Coker was a respected biologist at the Bureau of Fisheries who had been a close correspondent with Robert Cushman Murphy regarding Peru's guano industry. Coker's *This Great and Wide Sea* (1947) was a comprehensive introduction to marine science with sections on the history of oceanography, chemistry, physics, and life in the sea. In short, the text was a general introduction to oceanography—not exactly a work of science writing, but one that shared its intent.

Carson wanted to strike a balance between Lane's and Coker's accounts, but the creation of an oceancentric theme required more than genre bending. Carson's text emerged at a time when Americans found in the ocean a place to ease the psychological anxieties of the war, and Carson probably had some guidance along these lines. During the 1940s Carson began corresponding with other nature writers who were in the process of focusing their literature to ameliorate the psychic damage done by the war and the bureaucracy of modern life. For instance, she frequently joined Louis Halle Jr. on bird-watching walks along the C&O Canal towpath with the rest of the local Audubon Society chapter. Carson thought that Halle's popular *Spring in Washington* (1947) was the quintessential nature guide to Washington, D.C.'s environs, a recipe for curing the stifling bureaucracy of life in the capital. Another important literary contact was Edwin Teale, a name familiar to readers of *Nature Magazine* and other natural history journals. In late winter 1947 Teale and his wife started a seventeen-thousand-mile journey in South Florida and headed north with the spring. The resulting book attempted to chronicle the flow of spring as it moved northward and was a truly innovative piece of nature writing. At least for Teale, the journey was something of a balm for the terror of the war. "And while we waited," he explained, "the world changed and our lives changed with it. The spring trip was something we looked forward to during the terrible years of World War II."[41] Carson had the opportunity to lunch with both Teale and Halle, apparently to talk about developing a literary style. The advice they imparted to Carson is unknown, but we may gather from their work that they advised Carson to write of the ocean in a way that might have an ameliorative effect on a populace emerging from the throes of war.[42]

William Beebe continued to serve as another of Carson's frequent correspondents during the years she was engaged in writing *The Sea Around Us.* In the preface of the text, Carson admitted that her "absorption in the mystery and meaning of the sea have been stimulated and the writing of this book aided by the friendship and encouragement of William Beebe."[43] Beebe had high admiration for Carson's nature writing. She had received the honor of having a portion of *Under the Sea-Wind* included in Beebe's anthology of natural history writers, which included such notables as Aristotle, William Bartram, Henry David Thoreau, Charles Darwin, and John Muir.[44] In 1949 Beebe wrote a recommendation on Carson's behalf for the Eugene Saxton Memorial Fellowship, which she won in 1950. Apparently, Beebe also made a few personal calls to members of the selection committee. He even credited himself for seeing to it that Carson was awarded the Burroughs Award for nature writing in 1951.[45] It is difficult to trace exactly how Carson's writing was influenced by Beebe's, but she did respond to *Half Mile Down* with "a mixture of awe, envy, and gratitude that one of the two men who ever visited these depths was so exceptionally gifted with the ability to share those experiences with those of us less privileged."[46] Beebe's oceanic natural histories highlighted the inconsequential nature of humans when compared with the oceanic realm, a relationship that necessitated awe, respect, and humility.

The nature writing of Beebe, Halle, and Teale, as well as those of the biocentric writers mentioned earlier, informed the general theme of *The Sea Around Us.* Through a narrative of the ocean's history, Carson portrays humans as a species that emerged from an ocean to which they will return. She highlighted the fact that the earth was an ocean planet. Over vast periods of geologic time, the ocean had the power to weather down continents, and also the power to create life. Granted, humans have had the power to destroy through war and create through science, but human efforts pale in comparison to the transforming agencies of an omnipotent ocean.

This oceancentric theme infused the entire text, but it was the organizing principle of the first chapter. Carson began the book with a history of the ocean's development.[47] The first rains, Carson wrote, signaled the beginning of the dissolution of the continents. "It is an endless, inexorable process that has never stopped—the dissolving of the rocks, the leaching out of their contained minerals, the carrying of the rock fragments and dissolved minerals to the ocean. And over the eons of time, the sea has grown even more bitter with the salt of the continents." The destruction of the continents performed a vital function, for it was from those leached elements that the first life would develop. The sea "produced the result that neither the alchemists with their crucibles nor modern scientists in their laboratories have been able to achieve."[48]

Life slowly crept onto the shores of the continents and thus began the slow evolution of terrestrial life, the development of humans being a very small and recent part of a much larger evolutionary history. The earliest oceanic exploration symbolized the return of humans to the sea.

> And yet he has returned to his mother sea only on her own terms. He cannot control or change the ocean as, in his brief tenancy of earth, he has subdued and plundered the continents. In the artificial world of his cities and towns, he often forgets the true nature of his planet and the long vistas of its history, in which the existence of the race of men has occupied a mere moment of time. . . . [When on a long ocean voyage] he feels the loneliness of his earth in space. And then, as never on land, he knows the truth that his world is a water world, a planet dominated by its covering mantle of ocean, in which the continents are but transient intrusions of land above the surface of the all-encircling sea.[49]

Most reviewers took the moral exactly as Carson had intended. An Omaha writer reported that "Rachel Carson, in the *Sea Around Us,* places terrifying emphasis on man's helplessness against this enormous mass of water."[50] Others, like this Oklahoma reviewer, neglected the ocean altogether and homed in on the text's philosophical implications. "From one point of view, the penalty of such books as Hoyle's *The Nature of the Universe* and Carson's *Sea Around Us* is their impression of man with his own insignificance."[51] The moral was sometimes interpreted as an anodyne for atomic science. A New Haven reporter, commenting on Carson's book along with other oceanic natural histories by

James Dugan and Jacques Cousteau, thought that "the recent rash of books on . . . the sea are a result of the fact that man has found that there are fields of activity in which he never will emerge as the conqueror. Science has led him to the brink of disaster, and nature offers him a new vehicle for his irrepressible energies."[52] The moral could also be employed to reframe cold war anxieties. What impressed Bruce Barton most was

> the age of the world in contrast with man's brief span. . . . It seems to me that we Americans, in our thinking and planning, particularly regarding our so-called 'foreign policy,' tend too much to ignore the one fundamental that should never be ignored, time. . . . It can and will, with our help, eventually upset Communism. Provided that in our hurry for world salvation, we do not commit national suicide by draining our own land of its resources, and the veins of our sons of their blood.[53]

Yet another reviewer thought that *The Sea Around Us,* along with Maurice Herzog's *Annapurna* and Hemingway's *The Old Man and the Sea,* offered an enticing retreat from the complexities of civilization: "They offer us also vicarious courage, indomitable perseverance in the face of disheartening odds—qualities we know we need if we are to face and overcome the obstacles, the fears and the threats by which we all live surrounded today."[54] These reviewers found solace in *The Sea Around Us.* Whether they were concerned with the science of the nuclear bomb, the cold war, or the complexities of civilization, they all were attracted to the notion that the sea was somehow bigger than they were, completely out of their control. The ocean put human affairs into wider perspective.

More than a piece of nature writing that helped to resituate Americans' relationship to the natural world, *The Sea Around Us* was an extraordinary piece of science writing. The process of writing the book involved many trips to libraries and meticulous textual research. Carson's research was, needless to say, comprehensive. She began the project with a general outline that divided the project into three sections: a narrative of the history of the ocean's surface, its depths, the general morphology of its bottoms and basins; a description of the ocean dynamics of waves, winds, and currents; and a section on how the sea affects the lives of humans by controlling weather and providing natural resources.

Carson then drew from a wide range of scientific treatments of the sea. Most of the sources were natural histories of ocean geography, life, and dynamics published since the middle of the nineteenth century, but Carson also kept abreast of recent oceanographic developments, especially echolocation research on the seafloor and the scattering layer. A surprising amount of Carson's information came from nonscientific accounts found in newspapers and

Rachel Carson and Robert Hines examining shallow-water marine life in 1955. Courtesy of the U.S. Fish and Wildlife Service.

pilot guides, which Carson used to present exciting and dramatic descriptions of ocean islands and waves. Beyond textual research, Carson also interviewed and corresponded with naturalists and oceanographers. For example, she wrote to Thor Heyerdahl requesting a description of the surface life that he viewed from the raft of the *Kon-Tiki.* She discussed Pacific islands and the dynamics of waves and currents with Robert Cushman Murphy. Murphy also suggested that she contact Mary Sears and Maurice Ewing at the Woods Hole Oceanographic Institute. They were just two of the prominent wartime oceanographers that she interviewed; others included Henry Bigelow, a former director of Woods Hole Oceanographic Institute; Daniel Merriman of the Bingham Oceanographic Laboratory; and Harvard meteorologist Charles Brooks. *The Sea Around Us* was the product of the work of scientists and naturalists that had investigated the oceans, as well as Carson's own research. Her task was to organize and edit this material to create a publication that would introduce the sea to the reading public.

This work was thus an extension of her experience as an editor at the Fish and Wildlife Service, where her task was to help publish the work of government biologists in a readable fashion fit for public consumption. Two observations emerge from an examination of Carson's textual research. First, while she drew from oceanographic research conducted in the 1940s, much of her information came from sources of oceanography and oceanic natural history written before World War I. This understanding should come as something of a surprise, for *The Sea Around Us* is nothing short of a readable exposition of the oceanographic research that commenced during World War II. The timing of the two events are related, but not by cause and effect, as is often assumed by many historians.

Second, with an ever-vigilant eye for elegant writing, Carson took special note of the metaphors, phraseology, and style of modern and historical sources alike. Commentators on *The Sea Around Us* constantly call attention to the text's lyricism and poetry. However, this observation overlooks how Carson drew from early writings of natural history. One commentator recently noted that Carson's "literary skill is most evident in her ability to present complex scientific information with both clarity and grace." The claim seems innocent enough, though what the commentator means by "complex" is open to question.[55] For instance, few of Carson's notes speak to the many physical and chemical equations that were becoming increasingly prevalent in oceanographic research.[56] Carson once told her close friend Dorothy Freeman that "straight scientific exposition is not my 'contribution' to the world. It is, I agree, what you call lyricism. But if that lyricism has an unusual quality it is, I think because it springs from scientific fact and so rings true."[57] Because Carson was not a scientist, in the active researcher sense of the word, she came to "scientific facts" through scientific literature, much of which was written by natural historians so fond of lyrical representations.

The chapter entitled "Birth of an Island," for which Carson won the George Westinghouse Science Writing Award, is one of the true gems of *The Sea Around Us*. It was the first chapter that she systematically researched and drafted, and her agent, Marie Rodell, sent the chapter as a sample to two book publishers. At one point, Carson considered removing the material from the book to write a separate article, given the timely nature of the subject: the preservation of fragile island ecosystems in the war-torn Pacific. In a manner similar to Murphy's environmental histories, the narrative moves from a discussion of underwater geological formations, to the process of wave denudation, to colonization by organisms. This seamlessly leads to a treatment of the extraordinary species that evolve on ocean islands—species that have achieved a delicate balance within isolated environments. The chapter concludes with thoughts on the recent American colonization of the Pacific and a call to preserve these fragile islands.

The volcanic origins of sea islands is a paradoxical phenomenon, according to Carson, "in the ways of earth and sea that a process seemingly so destructive, so catastrophic in nature, can result in an act of creation."[58] Carson makes no direct reference to William Beebe's description of the eruption of a Galápagos volcano or his geologic interpretation of Bermuda, but the two certainly shared a cosmic wonder for such spectacles. Carson also draws from Beebe for his thoughts on the colonization of islands. By carefully examining organisms transported by air, water, birds, and seaweed, Beebe speculated on what kinds of animals and plants would soon inhabit an island if it had sprung up immediately in the middle of the Pacific Ocean. Carson, too, creates a similar hypothetical portrait.[59] She turns to Darwin in the Galápagos for evidence on the strange plants and animals that evolve on islands. She even quotes from his autobiography: "Both in space and time, we seem to be brought somewhat near to that great fact—that mystery of mysteries—the first appearance of new beings on earth."[60] Carson also drew from Ernst Mayr, Robert Cushman Murphy, and David Lack for their work on peculiar oceanic life. Her treatment of island conservation was directly a result of conversations with Robert Cushman Murphy, Harold Coolidge, and Raymond Fosberg, all of whom were heavily engaged in the preservation of fragile Pacific islands in the American Trust Territory of Micronesia, a huge swath of ocean and islands previously dominated by Japan. This research on Pacific flora and fauna caused Carson to echo recommendations of so many postwar naturalists that Pacific conservation and preservation should be an American priority. In her spiral notebook, she wrote that "fauna and flora conditions will never be nearer prehistoric conditions than now . . . [America needs] to preserve peculiar fauna and flora of Micronesia for future scientific study, use, and enjoyment."[61]

Carson's research on Pacific island life became the most overt conservationist message in *The Sea Around Us.* She summarized the problem with a sentiment that would become the ecological centerpiece of *Silent Spring:* "Most of man's habitual tampering with nature's balance by introducing exotic species has been done in ignorance of the fatal change of events that would follow." Oceania, she remarked, has experienced a long history of introduced species—rats, goats, snails, insects—that had detrimentally and irredeemably devastated fragile island ecosystems. She concludes the chapter by advocating the adoption of the principal motivation behind the Pacific conservation: "In a reasonable world men would have treated these islands as precious possessions, as natural museums filled with beautiful and curious works of creation."[62] In the end, "Birth of an Island" is a remarkable mixture of nature writing and science writing. Carson deftly weaves together the evolution of ocean islands and life and at the same time chronicles the work of naturalists (not oceanographers) engaged in Pacific exploration.

Carson's blend of nature writing and science writing drew from both modern oceanographic research and older natural histories of the ocean, many of which were elegantly written narratives and descriptions. The approach was deliberate. Whereas wartime oceanographers probed the depths of the ocean with the most technologically advanced equipment, Carson was drawn to a way of knowing nature that emphasized direct and personal observation of the ocean. Clearly the invention of the bathythermograph, used to create vertical temperature profiles, and sonar, used to map seabed topography and to investigate the enigmatic behavior of the deep scattering layer, equipped postwar oceanographers with unprecedented technologies in their quest to understand the sea. Although it is beyond doubt that Carson valued such technologically sophisticated research, she also felt that something of the human quest for understanding was being sacrificed. She concluded *The Sea Around Us* with a brief look at the modern *Sailing Directions and Coast Pilots* then being issued to all navigators.

The final chapter is as elegant as it is revealing: "In these writings [Coast Pilots] of the sea there is a pleasing blend of modernity and antiquity, with unmistakable touches by which we may trace their lineage back to the sailing directions of the [Norwegian] sagas." These writings directed navigators to carefully examine populations of sea fowl for clues on locating specific harbors and islands. They also suggested that mariners traveling through new waters seek out local knowledge. "In phrases like these," Carson wrote, "we get the feel of the unknown and the mysterious that never quite separates itself from the sea."[63] The moral demonstrates her deep sense of humility: humans will never completely divest the ocean of its secrets, and sometimes the old tried-and-true methods of knowing the ocean—direct experience and observation—served better than all the equipment of modern oceanography. But Carson also highlighted the *Pilots* because they were examples of good literature. At one point, she calls one of the guides "Conradian." A *New Republic* reviewer admired Carson's "appreciation of the writing of others; she makes use of many quotations, especially from the Pilot Books of the U.S. Hydrographic Office. . . . It is good to find someone valuing properly the literary merits of these publications."[64]

Edwin Teale thought that *The Sea Around Us* marked a new genre of writing that he called "science-literature," a work that simultaneously reported on the findings of science without sacrificing an intimate portrait of nature. Reviewers unanimously remarked that the text was an elegant combination of poetry and scientific fact. Graham Netting, assistant director at Carnegie Museum, also thought that *The Sea Around Us* broke new ground. What "passes as science writing today," he wrote, "is cold, thin gruel. What we obviously need is a new category of science classics, reserved zealously for those books that satisfy the twin criteria of imaginative research and literary craftsmanship."[65] More common were responses that praised Carson's lyrical voice. Articles resembling an *L.A. Times* review entitled, "Prose Sings Fascinating Sea Story," were commonplace in literary sections of newspapers and magazines around the country. One reviewer in the *Atlantic* was especially perceptive of what Carson had strategically achieved:

> I cannot vouch for the validity of Miss Carson's findings—or, rather, her synthesis of her and other oceanographers' findings—and possibly neither can the *New Yorker.* For, the story goes, when the editor of the *New Yorker* was congratulated on running the profile of the multifaceted ocean, he replied that he had asked a staff man to check Miss Carson's facts in the fact bin. And who had gotten the dope together in the fact bin but Miss Carson herself! The facts, though, sound right. And Miss Carson is scrupulously discerning between what is accepted as fact and what is proffered as hypothesis.[66]

Indeed, the facts did sound right. But just as with the earlier sentiment of biocentrism, Carson's oceancentric message rode on a wave of lyricism. Nature philosophy of this sort rarely comes through in what is typically considered scientific writing. Such ideas must be channeled through experience and literature. Carson imbued the messages of a biocentric point of view, one that worked against the grain of her vocation at the Fish and Wildlife Service. The idea was well worn by terrestrial writers, and Carson merely helped it to break through the continental boundaries and flow into the ocean.

But just as biocentrism emerged from the slag of industrial Manchester and New England, so too did oceancentrism require the proper cultural context: a war of death and destruction, a crisis in national anxiety, a scientific mobilization to wrest the secrets from the sea, and the chance to control it. Other than a simple translation of scientific fact into pleasant writing, *The Sea Around Us* functioned as an anodyne for a national consciousness struggling to come to terms with the all too dominant stature of the American ego.

THE MISSING CHAPTER

Given the trajectory of Carson's career as an environmentalist, *The Sea Around Us* seems to be something of an aberration. Carson's ocean was indomitable, unconquerable, mysterious. In many ways, Carson was continuing Beebe's project of portraying an oceanic sublime that rebuffs all efforts by humans to conquer, control, or destroy. Aside from the important topic of preserving Pacific islands, *The Sea Around Us* never entertained the possibility of humans altering the state of the ocean. Of course, there is a humble biocentrism here that is characteristic of nature writing in general, but there is little of the environmentalism that would emerge from Carson's pen some eleven years later. On the other hand, Carson may have expressed something of an environmental ethic, or an ocean ethic, not in what she put into *The Sea Around Us*, but rather what she left out.

It is odd that Carson, long a devoted employee of the Fish and Wildlife Service, chose to exclude any treatment of ocean fisheries. Indeed, during World War II, Carson had written a number of pamphlets for the USFWS that urged Americans to diversify their fish diet so as to alleviate wartime shortages of beef and to also ease the stress on soon to be overexploited fish populations.[67] As early as 1926, William Beebe had suggested that the human race might turn to plankton, "this larder of the ocean," as a food source.[68] But the possibility did not receive any serious attention until after the war, when scientists saw in the sea a new frontier of free natural resources. This was a critical time in American history, when the ocean was becoming a literal "new frontier."[69] In 1948 William Vogt and Fairfield Osborn published widely read books that raised the specter of environmental decay and its implications for human existence. Among the many topics covered, especially in the case of Vogt, was the alarming exponential growth of human populations and the inability of the earth to sustain its human residents. In Vogt's words, "By excessive breeding and abuse of the land mankind has backed itself into an ecological trap. By a lopsided use of applied science it has been living on promissory notes. Now, all over the world, the notes are falling due."[70] Many scientists mobilized to rectify the situation; others turned to the ocean in a vain search for another promissory note.

Shortly after the war Maurice Nelles, a research manager of the Allan Hancock Foundation, organized an expedition aboard the scientific ship *Velero IV* for the purpose of exploring the possibilities of harvesting ocean plankton. Two Carnegie Institution of Washington botanists began an effort in 1949 to harvest fresh and saltwater algae. A team of Berkeley marine biologists was also looking into marine farming. Yale professor Werner Bergmann was researching the possibility of desalinizing massive quantities of ocean water to turn deserts into productive land.[71] These optimistic projects were doused in the late 1940s when they proved technologically infeasible. Daniel Merriman, who Carson interviewed on the subject, and Gordon Riley predicted that ocean fishing could be increased fivefold with more efficient use of fishery technologies. "But to harvest any considerable fraction of the plankton of the world," Riley reported, "seems as fantastic as the old dream of extracting gold from sea water. By and large we must leave the plankton to the fishes."[72]

But Riley's opinion was not yet a matter of consensus. Carson was fully aware of the optimistism of the day and drafted a chapter, excerpted from the final draft, entitled "The Ocean and a Hungry World." Carson, referring to the criticisms of Vogt and Osborn, begins the chapter by outlining the current failings of agriculture to sustain the world's growing population and provides an elegant segue: "But from the plundered land we turn to the sea with many questions." She briefly describes the two schools of thought and comes down squarely on the pessimistic side of the issue. Annual fish catches could be doubled, but only at great expense. And while the earth's plankton amounted to an unfathomable biomass, no reasonable technology was available to retrieve this resource for human use. Scientists and conservationists, in Carson's estimation, should concentrate their efforts on increasing fish yields.[73]

But Carson removed the chapter. Moreover, there is no discussion of the issue whatsoever in the text. It is difficult to imagine the reason for the elimination of such a timely discussion. One historian has suggested that Carson realized that the chapter sounded more like the writing of an agent of the USFWS than a "curious scientist or reverent witness."[74] The chapter, she continues, lacks lyricism and is utilitarian in nature. If this was a hard-and-fast criterion for inclusion, then it is odd that Carson did not remove the chapter on the mineral content of the ocean, which is both utilitarian and, in my estimation at least, decidedly nonlyrical. The excision is troubling because Carson wrote the book to show the connections between humans and the ocean, a relationship that was becoming acute in the postwar period. Carson thought that the entire project was important because "the life that invaded the lands has already so despoiled them it is being driven back more and more to its dependence on the sea."[75] "The Ocean and a Hungry World" fits precisely into this rationale. Carson may have sensed a flagging enthusiasm for plankton processing as the difficulties of the project became apparent.

The removal is also troublesome given Carson's reactions to RKO's documentary adaptation of the book. Although *The Sea Around Us* (1952) won an Academy Award, Carson was completely dissatisfied with the final product. Irwin Allen, producer and director of the documentary, would have a more popular following in television with *Voyage to the Bottom of the Sea* (1964–1968) and *Lost in Space* (1965–1968) and the great disaster dramas *The Towering Inferno* (1974) and *Poseidon Adventure* (1972), but in the early 1950s, he merely solicited naturalists for stock footage of underwater scenes. Allen edited the footage into a film that Carson thought was "a cross between a believe-it-or-not and a breezy travelogue." In contrast, she had hoped for a film that possessed all the "beauty, the dignity, and the impressiveness of the Pare Lorenz script for *The River*."[76] The allusion to Lorenz is revealing, for *The River* (1938), while certainly providing an environmental critique of land use, was also highly optimistic that science and technology—especially through the example of the

TVA—could thoughtfully harness nature for social and economic benefit. This utilitarian message was very much at the heart of the missing chapter.

Given Carson's critique on the destruction of land, it is possible that she did not want to entertain the possibility of humans turning to the ocean as a panacea. She was also aware of the key ecological role played by plankton, and as she would later expand on this theme in *Silent Spring,* when humans tinkered with the lower levels of the food pyramid, the consequences were often unpredictable, and sometimes devastating. But these are just speculations. Aside from the Pacific islands, Carson's ocean remained inviolable.

Carson would change her opinion on the resiliency of the ocean some ten years later in a new preface to the second edition of *The Sea Around Us* (1961). In the intervening years, the Atomic Energy Commission had made a common practice of disposing of nuclear waste in deep-sea basins. Carson recalled her former attitude regarding the inability of humans to damage the ocean and provided a corrective. "This belief, unfortunately, has proved to be naive. In unlocking the secrets of the atom, modern man has found himself confronted with a frightening problem—what to do with the most dangerous materials that have ever existed in all the earth's history, the by-products of atomic fission."[77] Given that she wrote this new preface while frantically engaged in her *Silent Spring* project, it is not surprising that Carson warned her readers that irradiated plankton, even at low levels, would have magnifying effects in organisms higher up the food chain—organisms that were the fodder for the human race.[78]

Even though concern over the pollution of the oceans was still ten years away, Carson's oceanic nature and science writing did signify the development of a new relationship between nature, citizens, and the public, a relationship that would eventually become a characteristic component of modern environmentalism. The blinding success of *The Sea Around Us* led Carson to believe that she was witnessing a newfound appreciation of the natural world in the general populace. "I am convinced," she declared in her reception speech for the Burroughs Medal, "that we have been far too ready to assume that these people are indifferent to the world we know to be full of wonder. If they are indifferent, it is only because they have not been properly introduced to it—and perhaps that is in some measure our own fault." Nature writers are thus called to perform an important civic function: "It seems reasonable to believe . . . that the more clearly we can focus our attention on the wonders and realities of the universe about us the less taste we shall have for the destruction of our race."[79] Such was the new responsibility of environmentally minded nature writers like Carson. What could be said for nature writing could also be said for nature itself. At the reception honoring Carson as the recipient of the National Book Award, she gave her thoughts on the relationship between citizen and scientist:

> We live in a scientific age; yet by a strange paradox we behave as though knowledge of science is the prerogative of a small number of men, isolated and priestlike in their laboratories. This is not true. It cannot be true. The materials of science are the materials of life itself. Science is the what, the how, and the why of everything in our experience. It is part of the reality of living.[80]

Carson brought this message to her final book-length treatment of the ocean, *Edge of the Sea* (1955), which she described as "a seashore guide to the Atlantic coast done from what I think is a fresh point of view—much less a handbook for identification than something to give a glimpse of how life is lived among the wave-swept rocks and wet sand."[81] In the preface of the book she issued a directive for people to go out and examine seashore phenomena on their own: "Understanding comes only when, standing on a beach, we can sense the long rhythms of earth and sea that sculptured its land forms and produced the rock and sand of which it is composed."[82] The seeds of a grassroots environment movement were, in part, sown by naturalists like Carson and Beebe, who helped the American public realize their own potential as explorers and stewards of nature.

Among those who study nature writing, there is a kind of canon that goes a little like this: Henry David Thoreau, John Muir, Aldo Leopold, Rachel Carson. This is a narrative of about a hundred twenty years of environmental thought that slowly progressed from an antiurban pastoralism to a full-blown environmental movement. When Carson is attached to the end of that list, most scholars have in mind *Silent Spring,* and rightfully so. But if one were to break through the continental margins and head for open water, then the list would take on a different meaning. What Thoreau had done for New England forests and what John Muir had done for Yosemite Valley, Carson did for the ocean.

The Sea Around Us stands out as one of the most popular postwar treatments of the ocean. Carson echoed Murphy's message that human affairs were intricately tied to the ocean. She further elaborated Beebe's oceanic sublime by highlighting the insignificance of humans in the realm of nature. And she absorbed the biocentrism of a host of nature writers to convey a sense of awe, power, and mystery in the face of an almighty ocean. The ocean became valuable not as a resource but as a symbolic geography of humility. Such was the tradition of nature writers. The power of the text was bolstered by the scientific findings that Carson used to support her message. The postwar ocean was an environment fertile with potential meanings. Eugenie Clark embraced Carson's depiction of the sea as mother to all and put it to work in her domestication of the ocean.

CHAPTER FIVE

Eugenie Clark and Postwar Ocean Ichthyology

Gender, Oceanic Natural History, and the Domestication of the Ocean Frontier

After a thirty-odd-year career as an ocean ichthyologist, Eugenie Clark fulfilled a childhood dream when she participated in a fantastic deep-sea expedition that used the ever-developing technologies of ocean submersibles. Dubbed the "Beebe Project," National Geographic Society photographer Emory Kristof and Clark, then a marine biologist at the University of Maryland, traveled in the *Pisces VI*—a highly sophisticated deep-sea submersible—to a depth of two thousand feet off the coast of Bermuda during the mid-1980s. They used the submersible as a kind of deep-sea blind and simply waited to observe extraordinary life, especially the behavior of deep-sea sharks, a practice that was not all that different from what Beebe had done fifty years earlier. It is hard to overstate the connection to Beebe's project. The *Pisces VI* was equipped with lights, cameras, and bait. The naturalists patiently observed the ocean deep, and occasionally napped, until creatures came to them. Clark characterized her enthusiasm with her typical personal charm:

> When I was a child, William Beebe was my hero. . . . I told my family I would like to go down and be like William Beebe. They said maybe you can take up typing and get to be a secretary to William Beebe or someone like him. I said, I don't want to be anybody's secretary! I want to be like William Beebe going down . . . and I don't believe it, here I am doing just that . . . in the same place.[1]

Like so many youthful ocean scientists coming of age in the 1930s and 1940s, Beebe was clearly a touchstone, an inspiring icon worthy of emulation. Clark's autobiographical statement is just as important in revealing the constraints of women entering the realm of oceanic exploration. An administrative leader, a thoughtful naturalist, a pioneer of underwater science, and a popular portrayer of the ocean, Clark was, indeed, nobody's secretary, though she encountered these expectations from time to time. The "Shark Lady," her most popular moniker, was actually one of most important public figures in the domestication of the postwar ocean frontier in American culture.

The project of taming, controlling, or conquering the ocean can only be accomplished in the human imagination. But it has been a persistent goal of some naturalists, oceanographers, meteorologists, and technophiles for over a hundred fifty years. It is a story that opposes the representation of the ocean as an unbounded, chaotic, and unpredictable geography of wonder and mystery, one of the central motifs of Rachel Carson's *The Sea Around Us.* As the ocean continued to accumulate more and more credence as a literal frontier geography, it seems almost inevitable that Americans would believe themselves called to tame the ocean wilderness, a bit like Roy Chapman Andrews had done through the sport of the hunt. That the ocean provided an alternative geography similar to the West and the Great Plains is hardly exceptional. But such activities took place within unique cultural contexts. For Eugenie Clark, the domestication of the ocean took place within a postwar culture where the relationships between gender, science, and nature were in flux. Like so many popular ocean explorers, Clark's life was on constant display in American culture; some of these profiles highlight this peculiar mixture of the science of natural history, the dangers of exploring the ocean, and the feminine world of repressed sexuality and supposed domestic bliss.

In 1953 Clark was just beginning to enjoy the limelight as her best-selling *Lady with a Spear* (1953) garnered widespread attention. *Coronet* ran a pictorial essay entitled "Career Women" that began with a short retrospective: "A woman gave us radium; another helped unravel the secret of the atom. The hand that once rocked the cradle now blueprints sweeping skyscrapers or guides the scalpel in delicate surgery. Women everywhere have struck out on the paths to achievement." Included in the essay were engineer Beatrice A. Hicks, Pulitzer Prize–winning journalist Marguerite Higgins, radio manager Bernice Judis, industrial designer Maria Bergson, and academician and social advocate Mildred McAfee Horton. Two female scientists were also included on the list: Rachel Carson, author of the widely acclaimed *The Sea Around Us* (1951), who had "fallen hopelessly in love with the ocean as a child, and had been dedicated ever since to unfolding its mysteries," and ichthyologist Eugenie Clark. The short paragraph on Clark told of her recent expedition to the Red Sea: "there, in the shark-infested sea, a young woman was calmly spearfishing. . . . Dr. Clark has dodged sharks all over the world [and has] done much to broaden the frontiers of knowledge."[2]

Two years later, *Holiday Magazine* ran profiles on eleven "women in the world today." The series, which was worldwide in scope, attempted to "understand the triumphs, problems and defeats of a sex which has been often misunderstood, often terribly oppressed." Their profile of Clark flips back and forth between her "dangerous" work as an underwater naturalist and her ability to raise a happy family at the same time. "Who could guess," the article begins, "Eugenie Clark's story on first meeting her? She is small, pretty and unassuming. She has lovely hands, darkly sleek hair and a low, gentle voice. And yet Eugenie Clark would baffle a television panel trained in spotting rare occupations." The article then details some of her scientific work. She spends most of her time underwater goggling and spearing fish among "the shimmering beauties and dangers of salt-water reefs and seas." She encounters "dangerous sharks, barracuda and moray eels in the depths." She travels to remote areas of the world, lives in "tribal huts on tiny Pacific islands and has eaten raw squid and raw shark." The *Holiday* author attempts to balance this image of the heroic and adventurous naturalist by highlighting the domestic side of Clark. "It would be a mistake to think of Eugenie Clark either as a cold, totally scientific person or as one obsessed by one subject. Her marriage to Ilias Konstanitu is a happy, extremely close one and they have many pastimes, pleasures and friends in common." The two take enormous pride in their children, the article continues, but it has been a difficult year for the family because Ilias has been away from home for the greater part of every week. There is hope, for the family plans to relocate in Placida, Florida, where "the whole family will be together; . . . here the children will be able to grow up near the sea and later, in the sea with their mother. It is another happy ending for Eugenie Clark."[3]

By 1961 Clark was a well-known explorer, an intrepid skin diver, and a scholar of sharks. The *San Francisco Chronicle* ran a short biography and interview of Clark that coincided with a visit to the bay area. The piece was entitled "'Skin Dive,' Mother-to-Be Told," and reported that "there is only one phase of her skin diving that bothers Dr. Clark. She said she has been told by psychiatrists that this is a wholly masculine expression with erotic overtones," to which Clark responded, "But I really don't care. . . . If so, this is the only masculine thing about me."[4]

These three depictions call attention to the somewhat incongruous mixture of women and natural history. Clark was the subject of interviews and popular articles partly because she was engaged in the "dangerous" practice of exploration, thought to be a thoroughly masculine field of work. These profiles also attempt to come to grips with the working woman by emphasizing her domestic responsibilities; she could, at the same time, hold down a job and raise a happy family. These two dynamics are not unrelated. Clark doubly transgressed the social norms of her time by not only working, but also working in a traditionally male-dominated field.

Clark's history is illustrative of a culture of exploration in transition. Despite the many examples of women practitioners in natural history, early twentieth-century exploration was often considered a male activity. The Explorers Club, an elite class of urban explorers, did not list a woman's name on its membership roster until 1955. Indeed, a group of women travelers, geographers, and explorers, rebuffed by the Explorers Club, responded by founding the Society of Women Geographers in 1925.[5] The president of the Explorers Club, Roy Chapman Andrews, noted that he did "not see just where women fit into exploration. . . . I do not suppose that any man appreciates the feminine touch in most things more than I do. But on an expedition, how can a woman be anything but a liability? There are few women who are able to do technical or scientific work better than a man."[6] Of course, there was no dearth of women explorers in the early part of the twentieth century, but these naturalists had to consistently fight against the pervasive stereotype that the work of exploration could be practiced only by the "stronger" sex.[7] The gendered conception of exploration provided obstacles and opportunities that Clark negotiated within a postwar culture of domesticity.

More than dealing with a gendered tradition of exploration, Clark also had to deal with the problems faced by women within the broader structure of science. Despite the temporary move of women into academic positions during the war, the postwar period proved to be a major step backward in bringing more women into the scientific workplace. In 1949, for example, the Employment Opportunities Section of the Women's Bureau reported that women comprised less than 8 percent of the zoological positions between 1946 and 1947. One commentator noted in *Science* that "to discourage girls from going into science may be realistic because, as observers note, women are discriminated against in this field. They have less chance than men of being employed at their full potential, and they are employed at lower salaries."[8] One historian of science explains that this prejudice against women stemmed in part from a concern that female scientists were neglecting their domestic responsibilities.[9]

The obstacles for women making their way in exploration and science were mere reflections of wider social changes of the American workplace during and after the war. The story of women taking over their husbands' positions along assembly lines during World War II is well chronicled. If only for a short time, American women were encouraged, if not expected, to contribute to the war effort by exchanging apron for riveting gun. This seemingly dramatic social change is just as interesting for its themes of continuity. During the war, women were expected to express and elaborate their femininity even in the most male-dominated occupations.[10] They were to prepare the hearth for their husbands' triumphant return. When GIs returned home, American men resumed their old positions, and many women, but by no means all, returned to the household. Single women were encouraged to marry. A new culture of domesticity swept through white middle-class America, one that prized the virtues of reproduction, sexuality tamed within the contained space of the home, and the practice of raising healthy children—all activities located within the security of new federally subsidized suburban homes. The cultural matrix was thus prepared for the baby boom, and its attendant appraisals and prescriptions for behavior and sexuality: the Kinsey Report, Dr. Spock's *Commonsense Book of Baby and Child Care* (1946), and B. F. Skinner's behavior modification program.[11]

Themes of sexuality, fertility, gender, and behavior were at the center of an American middle class in rapid flux. But they were also key foci of Clark's ichthyological research program and worked their way into her representations of the ocean. And with good reason, for she seems to turn to ocean fishes, and here is the main point, in a way that justified her seldom-articulated brand of feminism that melded the postwar "career woman" and the homemaker. Clark fashioned her image as a scientist and a popular explorer in both her research and her popular writings by drawing from a postwar culture of domesticity that simultaneously made her status as a career woman and an explorer more acceptable. In Clark's capable hands, the ocean became an extension of the home, a safe place filled with innocuous and beautiful organisms. Clark's ocean was not the only natural geography to be domesticated in the postwar era, Disney's popular film series and television program *True Life Adventures* being the most obvious examples.[12] Clark was part of this wider culture of popular natural history, but more to the point, the goal is to explain *how* the domestication of the ocean functioned for Clark as a way of negotiating her own peculiar status as a career woman and a female ocean explorer.

AQUARIUM CULTURE: FROM PET TO RESEARCH ORGANISM

Clark's autobiographical account of her early interest in ichthyology tells of a seamless transition from childhood upbringing to naturalist. The narrative was written into Clark's most famous book, *Lady with a Spear* (1953), a personal exploration of a young naturalist working her way through New York, the distant Pacific, and the Red Sea. As a child, Clark's passion for swimming was encouraged by her mother and extended family, who taught her how to swim in the salty breakers of Long Island. Her affection for fish began with a childhood trip to the old aquarium at Battery Park, which later relocated to Coney Island, after which Clark filled her mother's apartment with aquariums containing fish, reptiles, and amphibians. Her work as a naturalist thus did not begin, as in so many other tales of naturalists, in the fields and wooded lots of her neighborhood. She was a child of New York City, and instead of venturing into the outdoors, perhaps the nature trails of nearby Ramapo State Park, Clark brought the outdoors to her little two-room apartment. She observed the feeding and mating behavior of her household pets. She watched "the mating of guppies, swordtails, and platyfish, which bear their young alive, and we saw them giving birth. It was fun to read and discuss all the aspects of these various modes of reproduction. It made it easier on those mother-daughter talks, too."[13] Clark was thus introduced to the mating of fishes at home with mother at her side. This experience with common aquarium fish and mating behavior extended into her training as a professional naturalist.

After completing an undergraduate degree in zoology at Hunter College, Clark began taking graduate courses at NYU. She had applied to Columbia University but received a cold reception from the chairman of the zoology department. Clark remembers him saying, "If you do finish you will probably get married, have a bunch of kids, and never do anything in science after we have invested our time and money in you." At the same time, World War II was well under way and, in her words, "industries were booming with war work and there was a shortage of men."[14] Wanting to simultaneously contribute to the war effort and secure funding for her graduate education, Clark turned to the want ads and found a job as a chemist at a plastic research lab in Newark. When not in the lab, Clark was studying under the American Museum of Natural History's curator of the department of fishes, Charles Breder, who sponsored Clark's master's research, a systematic and anatomical examination of the order Plectognathi (the famous poisonous blowfish of sashimi fame is a member of this order), with special emphasis on their puffing mechanisms.[15]

In 1947 Clark applied for and received a position with the Fish and Wildlife Service to study the fisheries in the Philippine Islands. According to Clark, when word got out that she was the only female scientist in the program, combined with the fact that her mother was Japanese, the FBI revoked her passport and sent her back to the states after a short stay in Hawaii. Perhaps with this incident in mind, she decided to continue her graduate studies at NYU. A Ph.D., according to Clark, was not completely necessary to pursue a career in zoology, but it was a helpful credential in a field where women had to deal with many other disadvantages. "Women scientists have to buck some difficulties when it comes to field work but I had one decided advantage. A man in my position often has a family to support and is not free to travel. I was independent and free to go anywhere and do anything I liked, and there was only my own neck to risk."[16]

For her dissertation research, Clark examined the reproductive behavior, primarily the evolutionary mechanism of sexual isolation, of platys and swordtails, the same kind of fishes that were in her first home aquarium.[17] Dr. Gordon, then the head of the department of animal behavior at the AMNH, had produced platy-swordtail hybrids that, although popular among aquarium enthusiasts, were born with a lethal black cancer. The two species often lived in close proximity to one another in shallow streams but seldom mated. So Clark set out to discover the mechanism that controlled such sexual isolation. The first part of the project entailed examining the precise nature of copulation under controlled conditions. This called for close observation—nose to the cold glass of aquarium in the museum greenhouse—and recording of interspecies copulation. Her friends jested, "What are you compiling there, the Kinsey report on fishes?"[18] In order to find the key moment of copulation, Clark used a micropipette to remove a sample from the ovaries for evidence of male sperm. She then used this technique in reverse to artificially inseminate female fishes with combinations of sperm from both platys and swordtails. Such artificial insemination techniques were new, and the success of the procedure was questionable.[19] In Clark's words, "I waited around like an expectant father." When one of her females produced twelve babies, she "felt like handing out cigars."[20]

So Clark's experimental research at the American Museum of Natural History, for which she was awarded a Ph.D. from NYU, extended her own childhood pastime, though with an added experimental approach that was common among animal behaviorists of the day. The kinds of questions she was asking of her research subjects reflected the wider social environment of the 1940s. At the same time that Clark went to work to fill wartime shortages of male employees, she was also manipulating the reproductive processes of aquarium fish, a procedure that she sometimes portrayed in masculine terms. Her experimental program paralleled an important social shift in the American workforce. At the same time, her research focused on factors such as reproduction and behavior modification that women were to foster during the war.

DOMESTICATING THE UNDERWATER WORLD

The experimental nature of her dissertation research meant that Clark spent only a modest amount of time conducting field research. By the time she had earned her degree, she had spent some time diving with popular naturalist Carl Hubbs in La Jolla; she had spent a week composing a key of Hawaiian plectognath fishes; she stayed at Woods Hole during the summer on several occasions; and she visited the AMNH Lerner Lab in Bimini. All were short excursions and never resulted in any sustained research or papers. This changed in 1948 when Harold Coolidge, executive secretary of the National Research Council's Pacific Science Board (PSB), visited the museum to inform the staff of fellowship opportunities for conducting research in the newly formed U.S. Trust Territory in the western Pacific. Clark submitted an application that drew on her expertise with Pletognathi and proposed a survey of the poisonous shorefishes in the Pacific. Coolidge had received advice from some of his colleagues that there could be considerable problems for a single woman in the South Pacific Islands.[21] Robert Cushman Murphy, a colleague in the American Museum, sat on the board that considered these applications, and he may have played some role in seeing to the acceptance of Clark's project.

The PSB project, called the Scientific Investigations of Micronesia, was a coordinated exploration of Micronesian anthropology, botany, zoology, and ecology. The U.S. navy's responsibility to administer civilian life in postwar Micronesia necessitated knowledge of island life in order to fit the practical need of conserving scarce natural resources for native use. The federal government therefore actively sponsored research in Micronesia to simultaneously transform the area into a militarized buffer region and make this form of colonialism as benign as possible.[22] Here is the all-important context for Clark's work in the Pacific. Undoubtedly, she was to compile a list of poisonous fishes—practical information for the American military. But what is more amazing was the way this peculiar style of gentle colonialism dovetailed with Clark's own domestication of the Pacific. In the decade before her visit, these islands had been the sites of carnage, death, and destruction, whereas Clark's Pacific is pacified, friendly, and homey. Clark's expedition to the Pacific resurrected a happy, domestic, and purely natural territory, then in the care of the U.S. navy.

The U.S. military provided much of the supporting infrastructure that made Clark's four-month expedition possible. After checking in at the office for the Scientific Investigations of Micronesia at the Bishop Museum in Honolulu, a military carrier whisked her off to the Kwajalein atoll en route to Guam. For two days she stayed at a navy base while waiting for repairs on the plane, and so she decided to put her time to good use. After visiting a native village made up of the rehabilitated natives moved from Bikini just before the 1946 atom bomb tests, she removed herself to the atoll's tidal pools, where she poisoned the water with a solution of rotenone, a method appropriated from the technique of scattering macerated *Derris* plants in bodies of confined water.

The method was perfect for securing large numbers of fish in short order. With pails of specimens in hand, Clark went back to the naval base's dispensary and proceeded to catalog and preserve the day's catch. A large group of sailors quickly gathered around Clark and assisted her work. They bombarded her with questions about the fish life that surrounded their little atoll, which led Clark to later reflect on the missed opportunities of American sailors in the Pacific who never bothered to look at their environment. "It made me wonder how less dull our boys would find their assignments on these islands if they knew more about the sea life surrounding them. . . . An inexpensive face mask or native-made goggles, bathing suit, and sneakers are all that are needed to see the coral reef world. What a thrill so many missed and how close they were to an endlessly absorbing pastime."[23]

Her next stop was Guam, which had long been an outpost for the U.S. military. She was received with all due hospitality, though she was carefully segregated from male personnel. She dined alone at a long table, attractively arranged with flowers. "Two Guamanian waiters attended me ceremoniously. . . . Every sip I took from [my glass] was immediately replaced by a waiter standing by with a pitcher. . . . When the smiling waiter was not filling my glass, he was shooing the flies away from me, scolding them in his broken English as if they were not Guamanian flies." The naval personnel arranged for native guides to escort Clark to Guam's most biologically rich reefs. After a long day of trap fishing with two native fishermen—a Mr. Quenga and his son, Ramon—the three went back to the latter's ramshackle house for dinner. The cooking was done on an open hearth, but "an ample supply of Cokes was kept in a huge, modern refrigerator which the average American housewife would envy. And they had a washing machine!" Americans, according to Clark, had no monopoly on those consumer products that defined the middle-class's standard of living. Quenga's family "can't be beat when it comes to all-out entertaining of a guest."[24]

Clark then moved on to Saipan, where she was housed in the luxurious abode of Commander Sheffield, the governor of the Marianas. The commander had relocated his family to Saipan and enjoyed all the comforts of western life. Clark was particularly interested in the Sheffield's peculiar pet, a large female fruit bat that Mrs. Sheffield took home and raised on milk, oranges, and, Clark noted, "most of the vegetable and fruit dishes that were part of the regular family meals. . . . To the Sheffields it was indeed a pet, and Mrs. Sheffield especially fondled and played with it as you would with a poodle." Her limited work in Saipan's tide pools took on a similarly domesticated flavor. Clark described a baby boxfish as if it had been "upholstered in a strip of cloth from the dress of a Ringling Brothers' clown." His face, she went on, looked like a Betty Boop caricature. "If one could keep this little fellow in a home aquarium along with guppies and angelfish, what an amazing pet he would be!"[25]

Her final destination was Koror in the Palaus Islands, where Clark used the newly founded Pacific War Memorial Station as a headquarters for the rest of her stay in the Pacific. It was in Koror that Clark's story takes its most important turn because it was here that she met Siakong, the native Palauan who taught her to spearfish. Siakong was employed by the Pacific War Memorial as a general handyman, and when he wasn't repairing fences and painting buildings, he was testing his meddle as "the best spearfisherman in the Palaus—or maybe in the whole world." It is hard to overstate Clark's admiration for Siakong. She described him as the epitome of masculinity, possessing the strength of some three native Palauans. Though a raggedly dressed fellow on land, "when he took these off to go into the water he was suddenly metamorphosed from a bum into a Greek God."[26]

Clark pays tribute to Siakong as more than a mere fisherman; she found him to be a deft natural historian of his local environment. After spending much time attempting to pull a triggerfish from its hiding place in a crevice of coral, for example, Clark marveled at Siakong's ability to quickly remove a similar specimen from its lair. The triggerfish, after moving headfirst into a hole, deployed a bony spine on its dorsal fin that locked the organism in place. A slight pressure on it, and the whole fin collapses. "Siakong of course knew this trick and had simply pushed the right button to back the fish out of its hole!" On any number of occasions, Clark would point out a desired specimen to Siakong, who would then go about the business of chasing the organism down. "He was a keen observer and his years of underwater experience made him an expert fish psychologist. He knew the ways of every one of the hundreds of varieties of reef fishes."[27] There was an unequal exchange taking place here. Clark would later receive international attention because of the success of *Lady with a Spear,* in part because of her depiction and appropriation of Siakong's primitive technology. In contrast, Siakong was to live out his life in relative poverty under the employ of a flimsy American infrastructure. Perhaps it was because, as a woman in science, Clark understood Siakong's awkward position and therefore praised his local knowledge and his skill as a natural historian, and thus eased this unequal exchange.

Clark seems to have been especially impressed with Siakong's primitiveness. Clearly drawing from the long entrenched myth that native peoples share a special relationship with the natural world, she often depicts Siakong as a seamless part of the Pacific fauna. On one occasion, "he dived and lay motionless on the reef, like an animal about to spring on its prey. His brown body and red loin cloth blended in with the kaleidoscope of colors on the surrounding reef." When she had to bail him out of jail for disorderly conduct, Clark found him "pacing the floor like a wild beast." The technology he used also highlighted his primitive nature. All of his spears were of "primitive hand type." He would have no part in the "fancy arbaletes of CO_2 cartridge spearguns" used by modern skin divers. When Siakong speared a fish, "his powerful arms were the only propelling force behind the spear." Siakong's spear functions as the linchpin that brings all of *Lady with a Spear* into context. It is an especially powerful symbol given that by 1953 Americans were all too familiar with pictures, movies, and stories of the damage that modern atomic weapons had heaped upon the Marshall Islands. Siakong's spear became something of a fetish when considered next to the awesome destructive power of nuclear science. And Clark's appropriation of this primitive technology functioned as an apology for America's presence in the Pacific; she pays honor and tribute to a culture that is in the midst of sweeping social change. She seized upon the image and the technology of the spear to fix the social development of Pacific peoples in a timeless past, a prelapsarian Paradise, a romantic Eden before the Fall—the Fall that was then taking place at the hands of American imperialism.

When Clark returned to the United States, she wrote up her notes on the poisonous fish of the Pacific, including Polynesian terminology, and submitted the report to the PSB, which then passed it to the navy. Whether or not any practical use came from this survey is still a mystery.[28] In 1950 Clark traveled to Egypt on a Fulbright to work at the Marine Biological Station at Ghardaqa in the Red Sea. In *Lady with a Spear,* the same sexual, domestic, and primitive themes that Clark had used to characterize the Pacific are relocated onto the dependent people and landscapes of northern Africa. Perhaps most interesting is her description of the pipefish: "The females, having deposited their eggs in the males long before . . . are free of maternal duties. The male is left with the responsibility of bringing the babies into the world." Sexual ambiguity, while an oddity, can be found throughout the natural world.[29] It is worth noting that Clark met Ilias Konstanitu in Egypt, and this became a critical part of the popular narrative; the two fell in love while Clark taught her future fiancé how to spearfish. The love story serves as an appropriate ending to this postwar travel narrative of a female naturalist. Her love for science, nature, adventure, and heroism seem all that much more acceptable when balanced with the story of her nuptial bliss.

Even before completing the manuscript for *Lady with a Spear,* Clark was moving into the sphere of public culture.[30] While in Egypt, she was asked to deliver a radio program on the reef fishes of the Red Sea. She began to articulate a message that was latent in her Pacific travel narrative but would become the central thrust of her future research on sharks. Responding to the question of whether she was afraid to dive around sharks and poisonous fishes, she remarked that once one becomes familiar with an environment, it is no more dangerous to explore the wilderness than it is to cross a busy city street. "It is one of the jobs of the marine biologist," Clark told her audience, "to make the environment of the sea more familiar and hence safer, through studying and understanding the animals which live in it."[31] Dispelling the myth of a dangerous underwater world, making the marine environment a safe place for human beings, was becoming a key plank in Clark's platform as a popular marine biologist.

Clark's Pacific and Red Sea fieldwork resulted in very few scientific publications. At the same time, these expeditions were among the most important events in her life. After her return to the States, she composed several travel narratives for *Natural History,* a publication of the American Museum.[32] Marie Rodell, the same literary agent that sponsored Carson's *The Sea Around Us* (1951), then assisted Clark in editing her hugely popular *Lady with a Spear* (1953).

AMERICA GOES UNDERWATER

A number of the book's reviewers found Clark's spearfishing escapades an extraordinary activity. Their wonder stemmed from Clark's participation in the predominantly male activity of diving, and specifically of spearfishing. The popularity of underwater recreation mushroomed shortly after the war. The explanation for why so many Americans took to the water defies simple explanation. One commentator suggested that it was the popular coverage of the United States Underwater Demolition Team, whose primary purpose was to dive down to shore and harbor bottoms to place and remove submarine detection nets.[33] Shortly after the war, teams of underwater soldiers were given the task of salvaging as much sunken equipment as possible. In these cases, the practice of underwater diving was a thoroughly militaristic, masculine, and athletic affair. These divers did not enjoy the luxury of scuba gear, but rather donned bulky and enormously heavy diving suits. One member of the navy's postwar salvage team noted that divers were "supposed to be the best physical specimens alive. They've all been some kind of tradesman before—like carpenters, welders, riggers or else athletes."[34] He also noted that every dive is a battle with nature. New commercial diving schools were established to train men yearning for an encounter with dangerous nature.[35] These were not glamorous jobs; they promised no recreational experience whatsoever. Work was always hazardous and required great strength. This was a man's world and a man's work.

In its initial surge, postwar underwater recreation had very little to do with technological innovation, for when coastal residents of California and Florida began entering the waters en masse in the early 1950s, they often did so with only a mask, snorkel, and fins. The technology for skin diving became easily accessible in the 1930s, but it was not until the 1950s that the activity began to become a sport. By 1954, California had over a hundred skin-diving clubs and boasted over a hundred thousand recreational divers; Florida possessed close to forty thousand regular skin divers.[36] In 1953 alone, over $40 million worth of underwater swimming equipment was sold by such outfitters as Abercrombie and Fitch.[37] The allure of underwater recreation was dominated by spearfishing, an athletic activity that tested the mettle of manly courage.

Readers of David Bradley's popular account of Operation Crossroads, *No Place to Hide* (1948), were introduced to spearfishing as Bradley hunted for specimens as evidence of radiation contamination of a Bikini reef.[38] Perhaps more influential was Hans Haas's account of his dangers and adventures of spearfishing off California waters. In *Diving to Adventure: The Daredevil Story of Hunters under the Sea* (1951), Haas told of his many dangerous encounters with moray eels, octopuses, and sharks of all varieties. Spearfishing thus held promise as an adventurous activity.[39] A *Collier's* account of "The Rover Boys under the Sea," a skin-diving club for military personnel on Wake Island, tells the story of an "indignant skin diver [who] punched [a] startled shark—and speared it. When a hammerhead shark bit another skin diver's catch in two off Hikureru Atoll in the South Pacific, he hit the shark with his spear gun, kicked it, punched it on the nose and kneed it. But when he tried to grab the shark by its jaw, he lost the tip of his thumb."[40] Two national organizations were established to encourage skin diving and promote yearly spearfishing competitions. As of 1953, the International Underwater Spearfishing Association held no events for women because "they have not taken to the rugged sport in such numbers as men."[41] In short, the underwater world became a playground for male hunters on the prowl. Clark's underwater adventures were thus a transgression into a predominantly male sphere.

By 1954, skin divers could be grouped into two general categories: those who entered the sea to hunt, and those who donned mask and snorkel simply to observe underwater life. In the latter case, skin diving became another effort to escape the daily grind, the here and now. One diver noted that "this sport is an adventurous escape from everything earthly and known—like a visit to the moon."[42] In striking contrast to the battle-against-nature scenario, conveyed by navy frogmen and implicit in the sea hunt, skin divers often felt as if underwater recreation served to bring humans into closer contact with nature. One diver noted on his first reef dive that "this is the domain of the fish and I feel like an invader but at the same time part of it."[43] We see here an important transition. The ocean depths were becoming a place of relaxed recreation; the ocean offered something other than sport and adventure. Skin diving united humans with nature; it was tranquil, calming, beautiful, even domestic. The innovation of scuba was the primary reason for this transition.

Scuba respirators were the products of wartime research, though none were actively used by Americans until after the war. They were sometimes compared to some of the other odious inventions of war research. "Out of the wealth of atom bombs, flame throwers, booby traps, and other World War II inventions, have come some devices that promise to survive and become indispensable in peace. . . . Like DDT and the jeep, these breathing machines will be of service to anyone who learns how to use them."[44] When nonmilitary divers began using scuba in the late 1940s, they were more likely to be making a living than recreating. The sponge and abalone industries changed overnight as scuba made diving for these shallow-water inhabitants much more expedient. Popular treatment of such divers, however, continued to highlight these activities as thoroughly dangerous affairs.[45]

Use of scuba for recreational purposes began in France, where Jacques Cousteau's respirator caught on very quickly. Mountaineers were among the first to see the potential of the new technology for transforming the underwater realm into a geography of sport. In 1946 Henri Broussard established the Sub-Marine Alpine Club in Cannes on the French Riviera. Members of the club emphasized hunting for reef fishes and setting records, but others might go down just to take a look around. An American traveler who visited Broussard quickly realized the therapeutic value of diving: "Perhaps these undersea caves would make perfect A-Bomb shelters, as well as retreats for asthma sufferers."[46] Despite the hyperbolic tone, scuba helped to transform the sea into a safe place, an anodyne to atomic warfare.

Scuba also transformed the work of Hans Haas. His early book on skin diving was a tale of bold adventures; danger lurked everywhere. In contrast, his popular film *Under the Red Sea* (1953), which made use of scuba, was almost comic in its depiction of somersaulting divers. In 1954, the American Museum's educational television program, *Adventure,* dedicated a show to the living ocean. Half the show was a viewing of *Under the Red Sea,* during which Clark provided a running monologue. Clark's gentle and becalming voice identifies species of corals and fish, and at the same time chuckles at the antics of the two divers playing in the weightless environment. Scuba helped make the undersea world a gentle place full of stimulating colors, delicate movements, friendly organisms, and playful fun.

To be clear, scuba did not completely erase the idea of a dangerous ocean. For instance, American scuba enthusiasts sometimes transformed the peculiar underwater experience into a very familiar frontier adventure. James Jones, author of *From Here to Eternity* (1951), quickly took to the sport which offered an exceptional challenge to those men seeking adventure in a "potentially dangerous situation." The ocean, for Jones was "the last frontier . . . [for] all those pioneer types who for whatever drives and reasons of their own need forever to keep pushing forward into dangerous terrain, with tanks on their backs instead of a horse underneath them, a face mask instead of a sombrero, a speargun in hand instead of a Colt's Frontier at the hip."[47]

When Clark entered Micronesian waters in the summer of 1949, even when her book was published in 1953, scuba was just beginning to work its magic in transforming the undersea world into a docile environment. As a spear-toting skin-diving enthusiast, when she dove into the ocean, she was entering an environment full of oceanic life and male spearfishermen. In the American consciousness, the ocean was still a geography of danger and mystery. But Clark's first undersea experiences were not those of dangerous encounters; she had noted the beautiful and romantic encounters with a tranquil seascape. Siakong, her native diving instructor, was represented as both a heroic hunter and a natural sea-dwelling creature who swam effortlessly among coral formations. He politely chided Clark's petty fears of innocuous sharks and barracudas. Spearfishing, Clark noted, "is the most pleasant method for the collector who enjoys swimming and observing fish in their natural habitat."[48] "Enjoyment" was the key characteristic of Clark's underwater explorations in Micronesia. She told a story of watching the amusing spectacle of Siakong playfully "bucking" a slowly moving sea turtle. The playful banter between the two underwater collaborators was always cheerful, pleasant, and even romantic. Though engaged in a scientific task, Clark's ocean was entirely recreational—not the man-against-nature variety, but more like a relaxing vacation.

Lady with a Spear was published shortly after her return to the States, and it met with much critical acclaim. It was a Book-of-the-Month Club selection, and Clifton Fadiman, the doyen of middlebrow culture, wrote a short review that favorably compared Clark's work with *The Old Man and the Sea, The Sea Around Us,* and *The Silent World.* He especially appreciated Clark's light and personal narration of her expeditions and claimed that her book was "really an account of a love affair with fish" encountered in a "magic fairyland or fairysea."[49] Gilbert Klingel thought that the book was "anything but a superficial adventure story." It provided "a series of sprightly pictures of the Pacific island spear-fishermen and Red Sea natives, and the result is a pleasantly readable . . . volume marked by a warm and youthful enthusiasm."[50] Lewis Gannet noted that it was slightly odd for such a lovely and unassuming woman to be visiting far-away areas engaged in such a risky field of exploration. "If all this sounds a bit exotic, let me assure you that, on the evidence of her book, Eugenie is what most of us like to think of as a typical American girl—matter-of-fact, down-to-earth, gregarious, athletic, fun-loving, and unashamedly curious about things which would certainly have made both her American and her Japanese-Scotch grandmothers blush."[51] A reviewer for the *Nation* suggested that "it is all related very simply and directly by a young lady who manages to seem typically American in everything except her career."[52] Although these reviews do show that readers were fascinated by Clark's depiction of an underwater "fairysea," they are more interesting for their attempts to make sense out of the author, a female naturalist engaged in a career somewhat at odds with her gender. Despite her participation in the exotic and male-dominated world of spearfishermen, Clark remained, in their eyes, a typical American girl.

The text so impressed Anne and William Vanderbilt that they donated a small coastal plot of their thirty-six-thousand-acre cattle ranch for the establishment of a marine biological laboratory on the west coast of Florida. Clark threw herself into the lab and quickly made it a premier site for marine biology research. Indeed, Clark became the same type of naturalist as William Beebe. Both were widely popular figures in charge of establishing research laboratories and field stations that were important sites for the scientific work of others. It was in this context that she made her greatest contribution in domesticating the ocean realm.

HERMAPHRODITIC GROUPERS AND GENTLE SHARKS: ICHTHYOLOGY AT CAPE HAZE LAB

Clark's work with sharks at Cape Haze Lab would earn her the nickname of Shark Lady. Clark readily embraced this name, but for very specific reasons. Indeed, Clark preferred not to tout her purely scientific work; she claimed that her most important contribution had been to help "dispel some of the myths about sharks that are so unfair to sharks."[53] Far from their popular image as the predacious thugs of the ocean, Clark's work revealed sharks to be "magnificent and misunderstood." But sharks were not on Clark's research agenda during the early days of Cape Haze. The quiet waters of Florida's Gulf Coast called for careful examination. Among the purely natural historical research that Clark continued to do at the new marine biological laboratory was an investigation of the functional hermaphroditism of a small grouper prevalent in coastal communities of western Florida. The sexual anatomy and behavior of *Serranus subligarius* was still something of a mystery in the mid-1950s, and Clark was confused by the presence of large colonies of these organisms, all of which had large bellies and thus appeared to be composed solely of females. In a popular account of her discovery, she recalled wondering, "Where were the males? After inspecting the reef many times, I learned to identify every fish I could find. There was no possible mate, even in the disguise of a different sex coloration or form. No mate! No source of sperm for the hundreds of unfertilized ovulated eggs inside each of the thousands of *Serranus* living in that reef."[54]

Clark brought the organism into the lab to conduct anatomical studies, and with the help of her AMNH colleague and teacher, Charles Breder, she learned that *Serranus* was a functioning hermaphrodite, possessing the sexual organs of both males and females. She wrote that "each adult individual can function as a male, a female, or both sexes simultaneously, depending on the situation."[55] After this anatomical discovery, Clark proceeded to observe the grouper's mating behavior in both the lab and in nature (the in situ observation marked Clark's early experience with scuba).[56] As with most other cases of hermaphroditism, self-fertilization is a rarity and is only useful in emergencies. Cross-fertilization is the general rule. Upon courtship, the grouper with the larger belly usually assumes the role of a female, while the other, manifesting an unusual temporary sexual dimorphism, produces banded vertical markings across its body. "The fish that chases is, as would be expected, in the male role." After mating, the sex roles may be reversed. Clark observed that during the "frustrating attempt at courtship, the fish begin to lunge and peck at each other; . . . usually the larger manages to force the other into a corner until it appears to 'give up'" and then becomes a banded male.[57] Mating then ensues. High-speed photography immediately after a mating revealed a further ambiguity. The unbanded female becomes banded for just a moment right after copulation. "At first this confused the otherwise clear-cut courtship relationship between an unbanded female phase and a banded male phase, until I realized that the banding on the leading fish is the *reverse,* or negative, of that of the male-phase fish."[58] Clark considered her work on the mating behavior of hermaphroditic groupers to be one of her most significant contributions to ichthyology. But at the same time she was conducting this research, another organism began to dominate the work of Cape Haze ichthyologists.

Clark was introduced to the possibilities of incorporating sharks into her research program by way of John Heller of the New England Institute of Medical Research. Heller was interested in shark liver as an anticarcinogen and called Clark for specimens; it was this inquiry that put Cape Haze Lab into the "shark-hunting business."[59] Other than researching the medical possibilities of sharks' visceral anatomy, Clark and visiting scientists also researched the function of the shark's abdominal pores, gestation periods, and size of brood. After having sacrificed thousands of sharks to exploring shark anatomy, she wrote that "the most interesting part of our work with sharks was studying the live

The LADY
and
The SHARKS

Eugenie Clark getting ready for a dive at Cape Haze Laboratory close to Saratoga, Florida, 1959. Reprinted from *The Saturday Evening Post* magazine, 1959 Saturday Evening Post Society.

animal, working with one individual day after day for long periods of time, and getting to know its personality."[60]

Always keen to ask and answer behavioral questions, Clark designed a research project that interrogated sharks' ability to learn simple tasks. On Heller's suggestion, Clark built a shark pen to maintain live specimens. In 1958 Dr. Lester Aronson, an expert in animal psychology, visited the lab and queried whether or not behavioral experiments had been conducted. He lectured Clark on the primitive nature of sharks; they possessed poorly developed brains and visual apparatus and were generally considered rather stupid. Clark later reconstructed the conversation: "'Besides,' [Aronson] said, 'they're difficult to keep as experimental animals, and no one has tried putting them into a Skinner box!'" Overhearing the exchange, "'Our sharks are smart, and boy, can they see us coming with the food!' Tommy said, defending his pets."[61] Although it is difficult to tell whether Tommy, a young boy who served as a labhand, thought of sharks as his pets, it is more telling that these are Clark's words. Here she is turning the ocean's inhabitants into a typical domestic phenomenon. After the pleasant exchange, Aronson then made a few suggestions regarding some simple behavior experiments.

Two lemon sharks and three nurse sharks became the organisms for a fairly straightforward behavioral experiment. Clark placed a white plywood target baited with a piece of mullet in the shark pen. She rigged the target with a bell that would ring when pushed with sufficient force. Over six weeks of training, the sharks were conditioned to associate both the target and the bell with food. After the sixth week, an empty target was placed in the water, and the sharks were rewarded with food after hitting the target.[62] In Clark's estimation, sharks were not as dull-witted as most ichthyologists had assumed. The domestic overtones of conditioning the top predator of the sea to answer a dinner bell are all too obvious. And the point wasn't lost on one reporter who wondered how to account for a shark that would "waggle its way into the shallows of the Cape Haze pen and beach itself to take food like a puppy."[63]

Some of the popular treatments of the shark conditioning experiments drew special attention to Clark's feminine characteristics:

> Three times a week a pretty, dark-haired, dark-eyed young woman walks briskly on a dock at Placida, Florida carrying a bucket of fish i n one hand and a notebook in the other. . . . She's going to feed her captive shark—but first the supposedly dull-witted beasts have to prove that they have learned how to order a meal. . . . The young woman, dressed in sport shirt and Bermuda shorts and looking something like a college student, is proving that sharks—one of the least understood of the world's beasts—can be trained to perform a sequence of acts.[64]

By the early 1960s dolphins were well on their way to becoming international stars in aquariums and on television. Clark seemed to be teaching similar tricks to sharks, with their vastly inferior nervous systems. The task was all the more amazing given that a pretty young woman who looked like a college student was the trainer.

One variant on the experiment proved to have tragic results. Clark, who assumed that sharks were color-blind, painted a target yellow instead of white. One of the lemon sharks approached the target and was so disturbed by the change that it did a backflip out of the water, and then refused to eat until it died three months later. Clark movingly recalled that she "felt terrible about his death. For more than a year, he had been a part of our daily activities. . . . We towed the remains of this once beautiful creature some miles out into the Gulf and watched it sink."[65] This particular lemon shark was much more than an experimental organism for Clark and her staff. It was a beloved member of the family whose death was treated with due ceremony.

These early behavioral experiments became the foundation of what became her recurring theme: sharks do not necessarily deserve a reputation of creatures to be dreaded and feared. The subject was particularly timely, given that the postwar ocean seemed to harbor life-threatening sharks under every swell, at least in the American imagination. The shark's reputation as dangerous predator was not new to the 1950s, but the issue received urgent attention because of America's activities in the Pacific theater during and after World War II. Given that many more Americans were entering the oceans as skin and scuba divers, both military and civilian, the navy called for a conference in 1958 to deal with the "shark problem."

Thirty-four participants from around the world came to New Orleans to present the latest work on the issue. Some papers dealt with the general taxonomy and natural history of *Elasmobranchii;* others dealt specifically with shark deterrents. Perry Gilbert, a Cornell professor of zoology, chair of the conference, and friend of Clark's, summarized the objectives of the conference:

> Why should countermeasures be considered? Man's interest in dominating his environment brings him into contact with sharks at four major impact points: 1. Sharks attack and maim or kill men under certain conditions. 2. Sharks inflect sever economic loss upon commercial fisheries . . . 3. Through psychological effects on man, sharks probably curtail man's recreation use of the sea. . . . 4. Sharks compete with man in the harvest of some of the more desirable food resources of the sea.[66]

Domination of the ocean, according to Gilbert, could be assisted through the development of chemical shark deterrents, exploding devices, meshing, fencing, and bubble hoses, but the conference's chief importance lay in highlighting the need for shark research of all kinds. The American Institute of Biological Sciences therefore created a Shark Research Panel that served as a clearinghouse for shark science. Under the continued support of the American military, the panel also created the now-famous shark attack file so that scientists could come up with some explanation as to why sharks occasionally attack humans. After only one year of work, the panel published its first report: the 1959 season had reported thirty-six provoked and three unprovoked attacks.[67]

In the published volume of the conference proceedings, Clark reported the shark behavior research she had been undertaking as well as some pointers on how to care for sharks in captivity. Clearly, this was an odd fit given the nature of the conference, but Clark's conquest of the shark menace had similar ends, though by radically different means. Many years later, however, Clark did make one important discovery in the area of chemical deterrents. She found some promise in an excretion of the Red Sea Moses sole, and although the chemical agent continues to enjoy a reputation as the only known relatively universal shark repellent, it has never been stabilized in a marketable form.[68] Justifiably, Clark was not overly optimistic; she pointed out that the desire for a repellent is more a psychological than a real comfort. "Your chances of being bitten by a shark," she wrote, "are much less than your chances of being in a crash when you drive your car."[69]

In 1959 Clark put one of her students, Sylvia Earle, in charge of Cape Haze and eventually secured a position at the University of Maryland. She took frequent field trips around the world, especially to the Red Sea, and began receiving the coveted patronage of the National Geographic Society. Although her research expanded to include various other oceanic organisms, she continued to assert that all sharks were not ferocious predators that posed immanent danger to human swimmers. Perhaps no publication communicated this message more forcefully than a *National Geographic* article relating her work on the sleeping sharks in the underwater caverns off the coast of the Yucatán Peninsula. Retelling an incident of coming face-to-face with one of these cavern-dwelling organisms, she wondered, "Why was this requiem shark—a member of the family that includes many of the principal man-eaters—on such good behavior? Why was it so unaggressive, so lethargic?" This portrayal of the docile requiem shark reinforced Clark's message that the trite conception of the predatory fish is horribly exaggerated.[70]

But sometimes the shark's reputation seems unflappable, for Clark's work on the requiem shark hit the newsstands just as Americans were flocking to bookstores and movie theaters for the latest installment of ocean terror. *Jaws*, Peter Benchley's fictional tale about the real phenomenon of seaside resorts torn between financial gain and ocean panic, probably had a greater effect than any other phenomena in stirring America's fear of the unknown terror that lies just under the surface of the ocean. In many ways, the ocean of *Jaws* carries similar mythological meanings as Herman Melville's. But the differences between *Jaws* and *Moby-Dick* also point to some of the new meanings and understandings of the ocean in modern American culture. The sperm whale of the mid-nineteenth century could no longer assume the monstrous role played by the great white in the 1970s. Peter Benchley came up with the story from memories of sharks while fishing for swordfish off Nantucket, bespeaking a simple economic change from whale fishery to swordfish fishery. More important, Benchley's great white shark stood in for all sharks—a common problem with the imagination—as an effective enemy because experiencing a shark in the ocean was an all too common affair for the many divers who were entering the ocean in the postwar period. The shark naturalist in the movie, played by Richard Dreyfuss, is a warm, sensitive, compassionate, and deeply human scientist; in a way, he is the male counterpart to Clark, a scientist who, in the estimation of one sociologist, corrected the image of the scientist in *Dr. Strangelove.*[71] Perhaps it is a testament to Clark's work on sharks that the popular press immediately sought out the expertise of less fictional biologists for confirmation on the film's portrayal of the white shark.

Clark actually enjoyed the movie and approved of it as a "good ichthyological science fiction." Although *Jaws* may have left the impression that all sharks were ferocious, it did some justice to the great white shark, a fish that, according to Clark, "deserves the reputation of being the most dangerous beast man can meet in the sea. No fancy or fiction can exaggerate the horror of its attack." But for the most part, the great white needed to be goaded with copious amounts of fish blood and slurry before attacking a human being. Clark generally followed her approval with an addendum: "Sharks can be trained to feed on cue and can be conditioned to press a target to obtain their food." In one review of the film, she alluded to the fact that most sharks are scared of humans; indeed, "man is more the predator of sharks than vice versa."[72]

Clark's most significant textual contribution in the demystification of sharks was an almost fifty-page article, abundantly illustrated with photographs from David Doubilet, that appeared in a 1981 *National Geographic.* "Sharks: Magnificent and Misunderstood" is the culmination of several years' worth of National Geographic Society–funded fieldwork while Clark was working at the University of Maryland. "No creature on earth has a worse, and perhaps less deserved, reputation than the shark," Clark explained. "Usually," she continued, they are "unaggressive and even timid toward man." She even invokes William Beebe's popular caricature of sharks as "chinless cowards." The article begins with a stunning description of a ride on a whale shark off the coast of Baja California. Clark used scuba gear to swim beside this toothless beast—the biggest fish in the oceans—and before falling behind, she grabbed onto the dorsal fin and "pulled up my knees and sat astride the shark's great back like a jockey." She enjoyed the ride for several minutes until the shark traveled too deep.

The article describes the new understanding of sharks that had begun to emerge since World War II because of extensive natural history research. She also describes her previous work at Cape Haze, and she brings her readers into the oceans off California, Florida, Japan, the Red Sea, and Australia, all to "contradict popular misconceptions of sharks as stupid, unpredictable eating machines, with nothing more than primitive brains and a good sense of smell." Even Clark's caged encounter with the great white shark left her "awestruck by their beauty." No surprise that she occasionally "reached through the bars and managed to stroke or pat a passing body."[73]

Despite the work of Beebe, Clark, and many other naturalists and scientists who attempted to teach the public that the dangers on land—for instance, while driving a car—are far greater than those encountered in the ocean, Americans continue to harbor a fear of sharks. Clark pondered the contradiction and provided the perfect explanation. The fear stems from our "familiarity with the rules and customs of life on land and our unfamiliarity with life under water." When humans enter the ocean, they enter the homes of sharks and other species. "We should mind our manners. If we break the social rules and antagonize our hosts, we should be prepared to deal with the consequences."[74] The ocean here is a home with rules and regulations, just like any other domestic space. Humans must enter this space and obey proper etiquette.

Clark was quick to point out some of the more severe violations of those social rules. "*Jaws*," she remarked, "was a fun and wonderful movie, . . . but it created this horrible reaction in people to go out and kill sharks."[75] In the summer of 1975 the number of participants at an annual New Jersey shark-hunting contest jumped fourfold compared with the year before. One shark biologist even claimed to have seen enthusiasts firing their rifles into the ocean. In the words of Glenn Frankel, captain of the *Blue Water*, a fishing charter out of Montauk, "that movie broke the [shark fishing] business wide open."[76] Whether for sport or for food, as was common in Japan, humans were levying a heavy toll on the shark's reputation. As early as the late 1960s, Clark, in articles, films, books, speeches, and interviews, called on humans to enter the ocean with respect and dignity for sharks and all ocean life. In many ways, the shark's place within American culture followed a similar trajectory to that of the wolf: from predatory pest in need of eradication to magnificent animal in need of protection and stewardship. Of course, such concern was a far cry from contemporary arguments regarding ecological roles; nevertheless Clark's was one of the earliest voices to aim this environmental ethic toward ocean inhabitants.

LIVING A DOUBLE LIFE: MOTHER AND SCIENTIST

Clark's research and popular writings on themes of sexuality, gender, hermaphroditism, behavior modification of sharks, and the proper behavior of humans entering the ocean all fit into the context of the cold war cult of domesticity. But Clark's personal life, far more than many of the other oceanic explorers that we have encountered, was a consistent subject of popular attention as well. So as a widely public figure, Clark had ample opportunity to express her own thoughts about human motherhood and the virtues of the ocean in creating healthy and happy families. Living next to the shore was the ideal environment for balancing her professional duties with her domestic duties. The point was in no way lost to a host of interviewers, who constantly attempted to reconcile the themes of her "double life" as a working mother.

Childbirth fit seamlessly into the economy of Clark's life as a marine biologist at Cape Haze Lab, or so it seemed. Drawing an entertaining analogy with her scientific preoccupation, she remarked that "in the midst of some of our shark-conditioning experiments, I took time off to have Nikolas Masatomo Konstantinu and even up the sex ratio in my family." But pregnancy did not interrupt her diving activities. Despite the concern of her male obstetrician, Clark continued to dive and swim almost up to the time of birth. She was even sure that "swimming activity throughout each pregnancy was a factor contributing to the extraordinary short and easy labor and birth of all four of my children."[77] The practice of skin or scuba diving was entirely compatible with the virtues of childbearing. Clark preached that diving had a therapeutic value for pregnant mothers under stress. She even suggested that when pregnant women feel "uncomfortable, just strap on an aqualung and go down to 16 feet. Then sit on the bottom for about an hour." Clark was also quick to highlight, in this same newspaper article, that scuba diving was crucial for her discovery of the hermaphroditic groupers.[78]

The virtues of raising a family in an oceanic environment were important to the health of the family unit, as well as in maintaining her career objectives. "Living and working beside and in the water somehow seemed to make childbearing and raising four small children easy to integrate with my job."[79] One of the primary reasons for achieving such a balance was that Clark shared the pleasure and burden of raising children with her extended family. "Hera and Aya enjoyed all the extras of having doting grandparents around, and I knew I could go on with work at the Lab, even with a third child on the way, since my mother was always available and pleased to help with baby-sitting."[80] She was just as certain that raising a family at Cape Haze provided a unique learning environment for her children. Clark wrote that the same anatomical discoveries that had been a daily part of Cape Haze research could be replicated while preparing dinner. The practice of cutting a fish open in the lab was not much different from cleaning and gutting a fish for dinner. And this simple domestic duty, according to Clark, was a perfect opportunity for enriching the minds of her children. "Children's first anatomy lessons should be in the kitchen while their mother cleans a fish or chicken. Packaged supermarket foods reduce this possibility, but fathers will probably never give up sport fishing and bringing home their catch."[81] More than teaching her own children the wonders of the sea, Clark began lecturing to small groups of visiting children, then to whole classes, and she eventually initiated a summer program where students could live and learn about the sea for an extended period of time.

Authors of popular articles were impressed with Clark's ability to reconcile work with family. An article profiling Clark in a 1960 issue of the *Doctor's Wife* entitled "At Home in the Ocean" shows how the simultaneous jobs of mother and underwater naturalist share the common trope of postwar domestication. The article begins with a quote from Clark: "'You've got to treat them with respect,' says Eugenie Clark, wife of orthopedic surgeon Ilias Konstantinu and mother of four. Eugenie, one of the world's leading ichthyologists, is not referring to her family but to man-eating sharks." The author goes on to describe the typical day of Clark's shark research program. First off, she checks on the "'guest' sharks who occupy a fenced-off area in Gasparilla Sound." Then she dives in the open water searching for new members for the "shark club." She then entices them to take a hook baited with "super market" meat. The article gives a few words on Clark's thoughts on the exaggerated nature of sharks' viciousness, and after the standard biography, culled from *Lady with a Spear*, the article concludes with a depiction of the Konstantinu family in which the ocean, again, becomes a place for familial bliss: "As each child came along, it was given an early dunking—all of them had a prenatal orientation since Eugenie, who believes swimming is good exercise for expectant mothers, kept up her sea diving until the last few weeks of each pregnancy. . . . The fascination of the world beneath the sea is indeed infectious—and from this mermaid mother, the whole Konstantinu family has caught the joys of being amphibious."[82] An article in *Sports Illustrated* explicitly noted Clark's "double life" as an ordinary housewife and as director of the Cape Haze Laboratory. The cover of the article shows Clark standing amid specimens of shark jaws that seem to be floating weightlessly throughout the room.

Eugenie Clark with family getting ready for a dive at Cape Haze. The popular press routinely called attention to Clark's family and husband. Reprinted from *The Saturday Evening Post* magazine, 1959 Saturday Evening Post Society.

> Dr. Eugenie sometimes finishes work with only a few strands of her dark hair askew. At other times she ends up smelling like a tubful of fish guts, but even on the grisly days, when evening comes, she washes the blood and Formalin away, tucks her fishiest thoughts well back in her mind and becomes the mother of four and the attractive companion of Dr. Ilias Konstantinu, an orthopedic surgeon who loves motorcars and mountains and is most tolerant of his wife's affection for the odd kingdom of the sea.[83]

Nevertheless, after seventeen years of marriage, the two separated. Clark left the Cape Haze Lab, moved to Buffalo, and married a philosopher, bringing the children with her. After a time, the children left to live with their father in Florida. These latter events elicited little comment from the press.

"In the beginning," Clark noted in a 1979 interview for *Ms. Magazine*, "I wanted to enter what was essentially a man's field. I wanted to prove I could do it." Whether it was the Columbia professor who tied Clark's gender to what he thought would be a certain lackluster scientific career, the government officials who denied Clark a passport to travel to the Philippines during World War II, or the director of Scripps Oceanographic Institute, Harold Sverdrup, who prohibited women from going on overnight oceanographic trips, Clark

Richard Meek's photograph of Eugenie Clark for 1965 *Sports Illustrated* article. The strings holding up the jaws were erased for the magazine. Courtesy of Time Inc. Sports Illustrated.

faced some significant challenges as a female scientist.[84] But Clark's feminism emerged much earlier as she broke from expectations about women and work, and women and science, as she edged her way into the male-dominated world of ocean exploration. Her status as a heroic explorer of strange lands and as a naturalist undertaking the "dangerous" activity of diving was mediated by themes of domesticity—gender, sexuality, mating, conditioning. One can't help but get the feeling that her work on guppies, groupers, and sharks was all somehow related to her own biography, which melded ocean exploration with the domestic overtones of the postwar family. Only by embracing a culture of domesticity, it seems, could she fight against the feminine mystique. A full-blown women's rights movement emerged in the 1960s that attempted to address discrimination in the workplace by, in part, fighting against the cult of domesticity that saturated American culture in the years after the war.[85] In contrast, Clark's peculiar form of feminism embraced a domestic ideology at the same time as it fought against prejudices against women in the workplace and women in science.[86]

In the end, this feminism played an important role in Clark's descriptions of a happy, safe, therapeutic, and domestic ocean. Whether it was her portrayal of the Pacific Ocean as a docile, friendly, and well-behaved playground, or her attempts to modify the behavior of sharks to answer a dinner bell, or her sermon on the therapeutic value of the sea for mothers to be, Clark created an oceanic realm that was an extension of the postwar domestic household. The popular coverage of Clark's work consistently balanced Clark the adventurer with Clark the mother. At this level, a domestic ideology made sense of a woman who had moved into a traditionally masculine field of work. Related to this, Clark drew from a culture of domesticity in order to legitimize her status as a naturalist who was doing the heroic and adventurous work of a male dominated science.[87] Her move into masculine territory was balanced by her status as mother and wife.

Less certain is the manner in which a domestic ideology informed her work as a scientist. Why did American Museum naturalists assign Clark the job of exploring the sexual behavior of guppies? Why the interest in modifying sharks' behavior? Is it significant that a naturalist who combined both masculine and feminine traits in an uneasy balance vigorously explored the hermaphroditic grouper? One answer to these questions is that Clark legitimized her postwar feminism by citing similar examples from the oceanic realm. As she moved into a postwar climate that discriminated against women in the workplace, as she fought for legitimacy in a power structure of science that discouraged female participation, and as she explored the perpetually "dangerous" seas that had previously been the work only of men of great daring, Clark drew from a culture of domesticity to ameliorate these transgressions. The final effect was to create a domestic, yet sometimes fragile, oceanic environment. This was a powerful message—a dramatic transformation in the concept of the oceanic frontier, a reconfiguration of space that dramatically changed the way Americans thought about and interacted with the ocean.

CHAPTER SIX

Technophobia and Technophilia in the Oceanic Commons

Thor Heyerdahl and Jacques Cousteau during the American Cold War

Americans' relationship with frontiers and technology may perhaps best be described as Janus-faced. For some, frontiers represent a geography that is merely away from the metropolis; they want to reunite with a "primitive" nature, to shed technology and civilization so that heart and mind can find new strength. Leave the telephone and the television behind and embrace a few days with wind and trees. It's an old idea that found a voice in Rousseau and the Romantics—a literary and artistic movement that can only be characterized as post-Enlightenment, a reaction to the dominance of scientific control and technological mechanization. But still others go to frontiers *with* technology in an attempt to exploit, improve, and transform a wilderness through technological innovation—to move hills, make water run against the force of gravity, turn nature into a productive machine, literally bring civilization to the frontier. The dynamics between these two rationalizations have thousands of permutations. They play out in multiple forums through time and space: bow and arrow versus rifle; dogsledding versus snowmobile; sail versus steam; live bait versus artificial lures; bike versus car; and so on. These two modes of thought and practice generally characterize the way many Americans interacted with the frontier West.[1] They also largely defined the way many Americans related to the ocean in the postwar era.

Technology mediates our interaction with frontier geographies, and the types of technologies that we use always speak to wider cultural issues.[2] This is especially true for the more inaccessible frontier geographies—high mountains, deserts, oceans, space. In postwar America, two non-American voices circulated through popular culture to articulate this twin dynamic as it related to the world's oceans. Thor Heyerdahl's popular *Kon-Tiki* (1950) was the best known of his many calls for the creation of a more "primitive" and benign interaction with nature and the ocean. His story is one filled with bold adventure and daring, but it is also one that works against the grain of modernity. In contrast, Jacques Cousteau became something of a technocrat of the ocean. Only through highly sophisticated diving, nautical, robotic, photographic, and engineering technologies could humanity truly begin to explore, understand, and take hold of the ocean frontier. There is something ironic in the fact that despite their many differences, both aided in laying the foundation for the articulation of an ocean ethic—an extension of environmentalism into the world's oceans.

To understand Heyerdahl's and Cousteau's interaction with and representations of the ocean, we must begin with some consideration of the cold war. This was a complex phenomenon in American history, a period that defies easy synthesis because it created an almost omnipresent pall that seeped into many facets of American culture.[3] The ocean, too, factored into this culture. The West was for the era of Manifest Destiny and frontier anxiety what the ocean (and outer space) was for the cold war. It was a world filled with threats, fears, and anxiety. Danger lurked everywhere: atomic and hydrogen bombs, communism, conspiracy, subversion, indoctrination, sexual deviancy, teenage rebellion, illicit drug use. Understandably, one of the primary goals of cold war culture was to contain those threats.[4] George Kennan, an influential American diplomat, articulated a foreign policy that emphasized the containment of the Soviet nemesis, and in many ways, containment of all anxiety became a concern for many Americans. One response to this perceived threat was to yoke the gods of science and technology into bolstering defensive and offensive capabilities in both economic and military arenas. It was a policy that championed the control of nature by engineering fiat. Another response was to search for an alternative, or at least some sort of temporary release, from the nerve-fraying anxiety of the cold war. Both options—control and release—represent two important ways that Americans were relating to the ocean during the cold war. They are encapsulated in the histories of Cousteau and Heyerdahl.

TECHNOLOGIES OF TERROR AND GEOGRAPHIES OF EDEN

It was truly one of the great adventure stories of the cold war era. In 1947 Heyerdahl and a crew of five left the coast of Peru on a "primitive" balsa raft similar to those used by pre-Columbian Andean peoples. Heyerdahl did not have a scientific degree. He was nevertheless resolute in his theory that South American Indians had crossed the Pacific to populate the Polynesian islands. As tenured ethnologists and archaeologists laughed at what they perceived as his dilettantism, Heyerdahl set sail to prove his theory correct. The mission captured the attention of the public eye. In the early 1950s, as evidenced by Heyerdahl's best-selling book about the trip, America had caught what several contemporaries referred to as Kon-Tiki fever. Indeed, *Kon-Tiki* graced best-seller lists for months and went through fifteen printings in the first eighteen months of its literary life. One critic called it "the greatest adventure story of our time" and marked it as the first text in a rush of adventure stories that were written in the early 1950s.[5] What was it about Heyerdahl's tale that held American readers captive? First we need some regional context. Heyerdahl's adventure was not just an oceanic tale; it was a story about the Pacific Ocean. How did Americans relate to the Pacific Ocean during and after World War II?

In modern western culture, the Pacific Ocean had long been considered an untouched and primeval paradise, a region of lush vegetation and simple people.[6] At the height of the Pacific conflict, new heroes emerged in the form of military construction battalions, the so-called Seabees, who leapt from island to island to support the soldiers on the front line. The heavy engines of rock crushers stood out in stark contrast against its island context.[7] Indeed, the machine age had arrived in the Pacific, an intrusion of the West that contrasted sharply with the stereotypical notion of the tropical Pacific. *Life* readers viewed such change through the eyes of war correspondent John Dos Passos. His first sight of an atoll in the Marshall Islands was of "the densely packed tents and Quonset huts around the edges of the field [that] have the look of an Arizona mining village. The bombing of our assault when we landed a year ago and the subsequent leveling of bulldozers chewing out airstrips has hardly left a tree standing."[8]

The end of the war left America with the charge of maintaining the peace in the Pacific. After some debate on the virtue of annexation, it was resolved by the United Nations that the United States would hold Micronesia in trusteeship. Most agreed that maintaining a military presence in the Pacific was a necessity. But the provisions of the trust were explicitly, if not disingenuously, anticolonial. America was to promote the social, intellectual, political, and economic self-sufficiency of Micronesia. The job fell to the navy, but in 1951 administration of the trust was transferred to the Department of the Interior. All throughout the period, debate centered on the acceptable limits of American occupation. It was clear that this was America's new empire, its new Pacific frontier.

Some commentators expressed an opportunist attitude. For instance, Warren Atherton, a former commander of the American Legion, remarked that at "the end of World War I, no opportunities were developed for American youngsters looking for new frontiers," but with World War II, "several million of our servicemen will return from the Pacific with a new knowledge of strange countries and their people. They'll say, 'me for the new lands of opportunity. I'm going back where I fought the Japs, and grow up with the country.'"[9] Atherton clearly gave new form to the classic frontier myth. "We must make thorough surveys and furnish aid to those who are qualified. . . . Clever Americans with bulldozers and tractors will open jungles and cut their way to mountaintops. Those with agricultural experience will bring new methods of making money from the soil."[10] Of course, many realized that most Pacific atolls would not support traditional modes of western agriculture. What is important here is the notion that the Pacific was a new American frontier ripe for conquest.

The colonial desire to turn the Pacific into an "American lake" contradicted the natural allure of the Pacific landscape.[11] A 1960 *Holiday* issue devoted solely to the Pacific described the totalizing effect of the war: "With a sudden painful jerk Oceania was swept into world history. Islands which had never seen an automobile suddenly became anchorages for huge aircraft carriers, bulldozers tore airstrips in the virgin jungle. . . . When it was over Oceania was part of the modern world. Her innocence was gone."[12] There is a certain remorse here, a nostalgia for a lost paradise, a regret over the sweeping power of war and modernity that raped a virginal landscape coded as feminine. Similarly, Laura Thompson, an anthropologist who had done extensive research in the Pacific, viewed the United States' trusteeship as a problem in conservation. She argued that island communities had developed self-regulating natural economies that were disrupted by the West. "Modern scientific discoveries and technologies, hand in hand with alien, exploitative ideologies, purposes, and habits, have to a large extent created the conservation problems. . . . During World War II and the postwar period the sudden impact of the outside world seriously disturbed the insular ecological balances."[13] Some of these imbalances were more disturbing than others.

Partly to perform one of history's most labor-intensive experiments, but just as much a drama of nuclear diplomacy, the U.S. military selected the Bikini atoll in their newfound territory as the first postwar site for nuclear testing.[14] Operation Crossroads, the name of the first postwar nuclear test, was an event designed for the world's eyes. News correspondents from the United States and countries the world over were invited to witness the spectacle. As such, the Bikini test received enormous attention in the popular press. Even before the test, commentators portended the possible end of the world. "One man feared gravity would be destroyed and everything would fall up," noted a *Newsweek* reporter. This commentator even added a reference to Revelation 8:8,9.[15]

Although most doubted that the Bikini test could cause the end of the world, many were not so optimistic regarding the fate of Bikini atoll. "How severely will the delicate balance of life, set by the hand of Nature be shaken?" wrote Edwin Neff in *Natural History.* "Centuries ago, guided by the light of the stars and the feel of the waves on the boat," noted a *Time Magazine* reporter, "the forefathers of the Bikinians had pushed their graceful canoes up the length of the Marshall Islands. Now graceless LSTs [tank landing ships] bore them to Rongerik, an atoll 140 miles to southeast."[16] The juxtaposition of the outrigger and naval vessels of war would be repeated in Heyerdahl's epic narrative.[17] Although many continued to see science as America's single hope to deter a communist aggressor, few could fail to recognize the disastrous implications of nuclear-age science. This ambiguity set the stage for the Kon-Tiki expedition, which purported to be not only an adventure of scientific exploration, but also a manifestation of "the desire to return to a primitive life that exists in most people—a move back to nature."[18] Heyerdahl's popular writings reflected America's search for a benign relationship with nature, a discourse that included secrets of the sea, mythic tales of ocean monsters, a search for the pristine history of humankind, and an attempt to demystify nature's mysteries with a kinder and gentler science.

It is important to appreciate Heyerdahl's Norwegian origin; indeed, part of his allure derived from his association with Nordic explorers. It was rare for any commentator to discuss Heyerdahl without mention of his Viking ancestors. Heyerdahl was born in 1914 in the small town of Stengaten, close to Larvik. He was only five when the family moved to Hornsjo, just outside of Oslo. Encouraged by his mother, who ran the Larvik Natural History Museum, Heyerdahl developed a penchant for collecting antiques, studying natural history, and visiting zoological museums at a very early age. He became entranced with stories of "primitive life." Heyerdahl related to his biographer how his distaste for modern civilization was cultivated during these early years. "Modern man," Heyerdahl told him, "had his brains stuffed full, not so much of his own experience as of opinions and impressions derived from books, newspapers, magazines, radio, and motion pictures. The result was an over-loaded brain and reduced powers of observation. Primitive man, on the other hand, was an extrovert with keen instincts."[19]

In 1933 Heyerdahl began his studies in zoology under the tutelage of a professor at the University of Oslo. He hoped to be taught about the living unity of nature and so was disenchanted with the college's emphasis on experimentation. Like Carson, he was willing to admit that dissection and the use of the microscope were necessary to understand the functions of organisms, but he felt that the scrutiny of intestines came at the expense of the knowledge of living habits and animal geography. In his opinion, the microscope in the laboratory should be supplemented by the telescope in the field.[20] It was a combination of his love for natural history and his mistrust of civilization that inspired him to continue his zoological studies in the Marquesas group of the South Pacific. He married Liv Coucheron Torp, a high school sweetheart who shared Heyerdahl's desire to return to nature, and together they set off in search for paradise.

The 1974 book that depicts their journey, *Fatu-Hiva,* begins with a reinterpretation of the Christian creation myth, an analysis of which helps make sense of Heyerdahl's proclamation to "return to nature." "Most Western religions," Heyerdahl wrote,

> uphold God's creation of both man and nature—nature in service of man. Living within an Edenic paradise, man did not need to alter paradise, for nature was perfect. God had provided for man with perfect efficiency. But man was not happy. Suddenly there was conflict between the Creator and the created. For, whereas God was pleased with his job, man was not. God was sure he had given man a perfect environment, an earthly Paradise. Man did not agree. While God rested [on the seventh day], man took over. Man wanted progress. Progress from Paradise. . . . What really bothered me was that adults said that God had created nature, yet they acted as if the devil was on their heels unless they severed all their ties with nature.[21]

The lapsarian moment here differs from the traditional notion of humanity's banishment from Eden by divine edict. Man believed, according to Heyerdahl, that nature needed to be altered. Man therefore banished himself from the Garden. Who is to blame? According to the western myth, we blame man's disobedience (and his thirst for knowledge), and then wonder why God delivered such a harsh punishment. Man should have felt remorse for his actions. But in Heyerdahl's reinterpretation, man took pride in his decision to progress beyond the constricting confines of Eden's gate. Progress, and science in the service of progress (simply and perhaps naively defined as altering the environment), have led to the modern predicament. The current situation was one in which culture and nature stood in isolation from one another.

Heyerdahl and Liv's expedition to Fatu-Hiva was therefore a return to prelapsarian nature, an attempt to heal the rift between man and nature, a turning back of the historical clock: "We'll go back thousands of years and start where our primeval ancestors left off. With our bare hands we will make our living and be one with nature. Then we can tell whether or not this modern world is as much a blessing as men think."[22] In 1934 the couple set off for their tropical honeymoon. During a six-week layover in Tahiti, they became good friends with Teriieroo, a native Polynesian who adopted the couple as children. In a farewell ceremony, Teriieroo renamed Thor and Liv with the native names Teraimateatantane and Teraimateatavahine—Mr. and Mrs. Blue Sky. With new names and new identities, the two set off on a schooner for Fatu-Hiva.[23]

Their experiment did not go smoothly. They got off to a gentle and possibly Edenic start. However, there are sometimes unforeseeable problems with relying too heavily on unmediated nature. Relations with native islanders were somewhat vexed. It seems that France's decolonization of the island instilled some enmity towards Europeans.[24] Thor and Liv contracted elephantiasis and made an emergency canoe trip to Tahiti over fifteen-foot swells. Before their hasty retreat from nature, Heyerdahl donned the anthropologist's hat and had learned from a native islander of a myth in which the ancient god, Tiki, had immigrated from the east on a balsa raft. Heyerdahl realized that Peruvians had a corresponding myth of their own in which an ancient leader, Kon-Tiki, was forced to set sail over the Pacific as a result of tribal warfare. When the swelling in his legs had subsided, Heyerdahl returned to Oslo, deposited the jars filled with South Pacific specimens in the natural history museum, and then turned his full attention to proving the South American migration to Polynesia.

THE *KON-TIKI* EXPEDITION: SIX NORWEGIAN SCIENTIST-ADVENTURERS

Despite his widespread popularity, Heyerdahl was usually a thorn in the side of professional anthropologists and archeologists. He had spent years, both before and after World War II, compiling botanical, zoological, archeological, and anthropological links between South America and Oceania. He pitched the manuscript to a number of American anthropologists, who greeted the theory with supreme indifference. Their key objection was that pre-Inca Peruvians did not possess the craft or skill to navigate the Pacific Ocean. Heyerdahl's suggestion that Native Americans had floated over on balsa rafts, for which there was some archeological evidence, met with similar disapproval. And so he set out to prove them wrong by conducting what social scientists call experimental archeology.[25] He would build a balsa raft, as per specification provided by historical evidence, and attempt to sail or float from South America to the Tuamotu Archipelago on the southern equatorial current.

As with all expeditions, Heyerdahl required support. The New York–based Explorers Club, whose members included Roy Chapman Andrews, Robert Cushman Murphy, William Beebe, and Eugenie Clark, provided some of the initial contacts. He received an inflatable rubber raft, radio, some scientific instruments, and other life-saving provisions from the equipment laboratory of the Air Material Command and the U.S. War Department. A British medical officer even provided "a mysterious shark powder" to be tested for its shark-deterrent capabilities.[26] In return for this bounty, which Heyerdahl insisted would not unduly influence the historical experiment, the expedition members would provide testimony on how the equipment performed. Heyerdahl selected five Scandinavian colleagues to accompany him on the trip. All were selected for their distinctive skills; several were fellow veterans from World War II. When the questions of money, crew, and emergency provisions had been settled early in 1947, he flew to the steamy interior of Ecuador to begin preparing the raft.

Fate threw a barrage of complications in Heyerdahl's path. The project of extracting suitable balsa logs from the Peruvian jungles during the rainy months of the South American summer provided an exciting narrative of ethnographic and natural contact. In the end, it fell to a battery of "half-breed" native Americans working under the auspices of a wealthy plantation owner to locate and secure the logs, each of which was given a name according to Polynesian tradition: Ku, Kane, Kama, Ilo, Mauri, Ra, Rangi, Papa, Taranga, Kura, Kukara, and Hiti. The logs were floated and then transported by steam paddle ship to the naval yard in Lima. President Bustamante y Rivero, six presidents removed from the administration that expedited Robert Cushman Murphy's exploration of the Peruvian guano islands, approved Heyerdahl's request for assistance in constructing the raft. The site of construction was a microcosm of *Kon-Tiki*'s wider cultural significance. Heyerdahl noted, "It was really a pathetic sight. Fresh-cut round logs, yellow bamboos, reeds, and green banana leaves lay in a heap, our building materials, in between rows of threatening gray submarines and destroyers. . . . The descendants of the Incas have moved with the times; like us, they have creases in their trousers and are safely protected by the guns of their naval craft. Bamboo and balsa belonged to the primitive past; here, too, life is marching on—to armor and steel."[27]

Steel and transuranic technologies of death and destruction, as well as speed and efficiency, littered the postwar oceanic environment. But for Heyerdahl, the rejection of modern technology was a question of scientific objectivity. He noted in a *New York Times* article, just before the start of the expedition, that "there are no nails or plugs in the raft"; the raft was constructed exactly as the pre-Columbian Americans had done so fifteen hundred years ago. The scientists had to strip themselves bare of modern conveniences because they needed to be "equipped as nearly as possible in their [the Peruvian sailors] way."[28] Needless to say, Heyerdahl found great wisdom in the native design. "Among primitive races," Heyerdahl noted in his Academy Award–winning

Thor Heyerdahl (second from the right) and other explorers planning an expedition at the New York Explorer's Club, 1946. Patrons for adventures could usually be found at the Explorer's Club, an elite institution dating back to 1904. Courtesy of the Kon-Tiki Museum.

documentary *The Kon-Tiki Adventure* (1953), "the golden rule was: don't resist nature, but yield to her commands and accommodate." It seems that the use of military-designed life-saving technologies would be excused without prejudice. On April 27, 1947, Gerd Vold, the expedition's secretary, christened the *Kon-Tiki* with a coconut amid a large diplomatic send-off. The raft was towed out of Callao Harbor by naval tug and cast adrift into the Humboldt Current.

For a hundred and one days, the *Kon-Tiki* traveled north and west riding with prevailing trade winds and ocean currents—over four thousand three hundred nautical miles. The *New York Times* continued to provide updates of

Picture of *Kon-Tiki* under sail during a large swell. A crew of six floated on the balsa raft for a hundred and one days in 1947. Courtesy of the Kon-Tiki Museum.

the raft's progress; reports that were made by one of the crew who passed along brief updates via ham radio. The world heard very little about the voyage, other than their position and a few details on how the raft was holding up. On board, the first order of business was to master the craft itself. Their only navigational priority was to keep the stern into the wind, but even this task required some dexterity and experience. They ate some provisions from the mainland, but the larger share of their diet came from the sea itself. Indeed, floating objects like the *Kon-Tiki* often attract a profusion of ocean life, and their encounters with whales, dolphins, flying fish, and sharks seemed to keep them well fed and not a little entertained.

The abundant marine fauna afforded them opportunities for scientific investigation as well. Quite by accident, a specimen of *Gempylus,* or snake mackerel, found its way, along with many other wayward organisms, onto the bamboo mat of the raft. Heyerdahl reported that this was the first living specimen of the fish ever obtained, and he put the discovery into the context of a subtle critique of modernity: "The sea contains many surprises for him who has his floor on a level with the surface and drifts along slowly and noiselessly. . . . We usually plow across [the sea] with roaring engines and piston strokes, with the water foaming round our bow. Then we come back and say that there is nothing to see far out on the ocean."[29] Just as he was proving that a balsa raft was an appropriate piece of technology for transporting humans, so too did the practice of drifting on ocean currents provide a setting for inquiry into the natural secrets of the ocean. "One has to voyage on a prehistoric vessel," Heyerdahl noted in his documentary, "in order to discover new zoological species in the twentieth century." Rachel Carson was so impressed with this mode of scientific inquiry that she requested a description of ocean life from Heyerdahl. He responded kindly by letter, and almost the entire correspondence was printed in *The Sea Around Us.*

At journey's end, the *Kon-Tiki* foundered on the barrier reef of an uninhabited island in the Tuamoto Islands. The crew towed the craft and all their belongings onto the island with the help of several Polynesians who lived nearby. Contact was quickly made with the proper authorities, and the six Scandinavian adventurers moved into celebration mode. A *New York Times* editorial marked the safe landing as the end of "one of the great adventure stories of our time. It was a voyage at once into the unknown present and the remote past." The editorial described the scientific implications of the mission's success. Indeed, there might be alternative explanations to armchair anthropologists who were bent on Asian diffusion. But the writer concluded by waxing on the wider cultural meaning of the expedition. "Otherwise it was a voyage many of us might envy, a happy holiday of sun-drenched days and starlit nights far from the maddening crowd and all the troubles of our bewildered modern world."[30]

After the parties, parades, medals, and honors came the business of lectures and publications. Shortly after returning stateside, Heyerdahl pondered the significance of the expedition in an article entitled, "Six Men on a Raft: A Tale for Statesmen." The point of the article was to show how an oceanic adventure aboard a primitive raft renews and purifies human relations. Such voyages were the antidote for the volatility of postwar foreign relations, and Heyerdahl recommended that "the leading statesmen of the world should be put aboard a raft and sent to sea for a few months."[31] The purifying effect of such a prospect would certainly be a result of teamwork under trying circumstances, but Heyerdahl was promising more. He not only wanted to turn back the clock to explore the possibility of a migration theory, but he also wanted to help humanity reunite with nature as a predicate for peace. "Our voyage had, in fact, united us with nature. We had developed the senses of primitive man. We felt as if the ocean wind and the salt water had washed through body and soul and freed us from the problems that beset civilized man."[32]

KON-TIKI FEVER

Heyerdahl's rafting adventure might have been little more than an odd madcap event if it had not been for his runaway best-selling account of the expedition, published by Rand McNally in 1950, and a feature documentary distributed in 1953. The book was a different creature when compared with his *American Indians in the Pacific* (1952), the author's preexpedition manuscript outlining his theory of Native American migration. *Kon-Tiki* was crafted for a popular audience, though Heyerdahl never seemed to pander to a readership bent on mindless adventure. Although anthropologists and archeologists alike had more than a few misgivings with Heyerdahl's theory, general American readers were far more forgiving. The book remained on the *New York Times* best-seller list for nearly seven months. Reviews in the popular press were generally favorable, and the book's success elicited extensive commentary. What did *Kon-Tiki*'s sizable following signify?

Advertisers for *Kon-Tiki* publisher Rand McNally touted the significance of the book's success and provided some interesting cultural insights in a newspaper advertisement:

> Is it just coincidence that four of the top national best-sellers in fiction and non-fiction have the sea as their central theme? Two of them—*The Caine Mutiny* and *The Cruel Sea*—are terrific stories, human dramas not merely acted out against the sea as a stage setting, but with the sea itself playing a central role which limits and controls the actions of the men who find themselves on it, or in it. *The Sea Around Us* is the fascinating and unusual history of the life—organic and inorganic—that has gone on in, on, and under the sea for ages and aeons past. And by now everybody knows that *Kon-Tiki* is the thrilling saga of six men. . . . We do not claim that *Kon-Tiki* started a "trend"—but it certainly has served to remind readers everywhere that the ocean (like the mountains) is one of man's last repositories of mystery, and that a great tale of the sea still has power to hold "children from play, and old men from the chimney corner."[33]

Rare advertising copy indeed. If the advertisers were effective in highlighting the sea as a "repository of mystery," they were less so in explaining the cultural longing for that mystery.

Perhaps it was *Kon-Tiki*'s critique of modern science and technology. For instance, James Michener, the dean of postwar travel literature, noted that the crew "came to know and love the great ocean as few have ever done." What was the epistemic nature of this knowledge for Michener? The review begins by calling attention to the scientific theories that Heyerdahl set out to prove. But if science was the way in which Heyerdahl came to know the great ocean, then it was a science whose value lay in knowing the ocean through direct experience as "they came to know every mood of the ocean and wrested away many secrets." Michener goes on to criticize Heyerdahl's science as well as his literary

Erik Hesselberg, Heyerdahl's navigator, playing guitar on the reed mat of the *Kon-Tiki.* The adventure spurred a profusion of tiki popular culture in restaurants, bars, and hotels across the American sunbelt. Courtesy of the Kon-Tiki Museum.

style. Both fall short a grade. But the book was important because it was a "vivid account of sea and raft and great arching sky."[34]

More than just *Kon-Tiki*'s importance as a treatise on the ocean, Michener esteemed the crew's courage. "This is a book to make one proud that we still have in the world six young men who would venture upon the ocean on a raft, merely to prove an idea. It is good to know that such courage still exists."[35] Michener fails to note the social nature of scientific authority that the story also invokes. For there was something subversive in Heyerdahl's exploit. An amateur anthropologist was attacking the established theories and traditions of the professional academy. The power of *Kon-Tiki* lies, in part, in this underdog narrative. When the expert scientists would not listen to Heyerdahl, he merely proved his theory by living it. Nothing could be more tangible and so straightforward. Heyerdahl's science provided an alluring alternative to esoteric, experimental, and atomic science of the cold war. *Kon-Tiki*'s readers, just as Michener, admired Heyerdahl and the crew for their courage to face the duel adversity of the Pacific Ocean and the structural power of ivory tower scientists who were busy destroying Pacific atolls with incomprehensible technology.

Kon-Tiki was more than just a good read; it subtly wove its way into American culture. One of Heyerdahl's biographers characterized the profusion of postwar Polynesian pop culture as a kind of Kon-Tiki fever that manifested itself in consumer culture's appropriation of the Tiki image. Drinks, chocolate, candy, perfume, paints, cookies, butter, sardines, leather goods, porcelain, silverware, building sets, souvenirs, matches, insecticides, calendars, ornaments, cartoons, music, and board games all embraced the Tiki image in order to capitalize on its poignant cultural power.[36] The power of Tiki-themed consumer culture reinforced other trends in America's interaction with the postwar ocean.

The primitive and benign technology that mediated Heyerdahl's engagement with the ocean percolated through an American popular culture beset with cold war anxieties. Perhaps the most obvious link to Heyerdahl was the profusion of Polynesian-themed restaurants and bars that mushroomed through America's sunbelts and major cities in the 1950s. The paragon of all South Seas cuisine, even before the war, could be found in Oakland, California, at the campy trading post known as Trader Vic's. The restaurant, pioneered by the peg-legged rum expert, Victor Bergeron, was established in 1937. Lucius Beebe called it an "Oakland institution," an exotic environment filled with "obsolete anchors, coiled hawsers, . . . fish nets, stuffed sharks, capstans, long boats, ships bells, Hawaiian ceremonial costumes, tribal drums, boathooks, and small-bore cannons." Trader Vic's offered a line of fruit and rum cocktails and a menu of Pacific-rim cuisine adapted to American taste, attempting to transport its patrons to distant Pacific locales for a reunion with a more natural environment.[37] In 1946 Bergeron packaged his recipes in a cookbook that was a runaway best seller. "Pursuing my interest in the South Seas," he wrote, "I've discovered that those gentle natives, the Polynesians, know how to have fun in simple, unaffected ways. The beauty of their surroundings, the ease with which they acquire food and clothing are not for us who have built up a complicated, war-torn civilization which is a far cry from island ways. But some of those ways can be injected into our own lives, to help us relax and have fun."[38]

Tiki culture moved beyond Trader Vic's to inspire fashion design, like the lava-lava wrap, as well as hotels and theme parks, like Disneyland's animatronic Tiki Room. Polynesian theme bars, restaurants, hotels, and Disney exhibits provided middle-class Americans with a vicarious engagement with ocean environments. For those who were only slightly more adventurous, cheap airline fares enabled a more direct, though only slightly less contrived, experience with ocean islands. Pacific vacations had long been a luxury mostly enjoyed by America's elite. Indeed, from the late nineteenth and through the first half of the twentieth century, the Hawaiian seascape played host to a prestigious list of America's and Europe's financial and cultural elite. After World War II, Hawaii restructured its economy around tourism, but now for middle-class mass consumers.[39] By 1960 Pan Am Airline offered ten thousand seats per week to Hawaii; throughout the 1960s hotels increasingly brought tourists to Hawaii on packages designed specifically for middle-class Americans and their need for a vacation to escape the anxieties of cold war life.[40] What Pullman Palace cars and later the automobile did for tourism in the West, the airplane did for Hawaii and other oceanic locales.[41]

The Tiki themes that resonated in the explosion of postwar tourism in the Pacific intersected with one of the clearest markers of Tiki culture's pervasive spread in middle-class American life. Surfing was an activity that was woven into the cultural life of pre-European-contact Hawaiians. Late nineteenth-century missionaries, sailors, and pineapple plantation owners, the harbingers of official U.S. imperial policy that annexed the islands in 1898, touched Hawaiian culture at its core. The withering impulse of American hegemony brought a quick end to the activity, but surfing was then resurrected in the interest of early twentieth-century tourism, a period in which the body of surfing and swimming sensation, Duke Kahanamoku, was given the task of advertising Hawaii as a "primitive" and "exotic" paradise.[42]

There were only a handful of American surfers in the first half of the twentieth century, but the sport grew quickly in the late 1950s and early 1960s because of technological advances as well as the rush of Hollywood surfing films, which began with Columbia Pictures' version of Frederick Kohner's *Gidget* stories, and the music of the Beach Boys.[43] These explanations overlap in meaningful ways with the sport's social profile; American surfers were generally white young men enjoying the fruits of postwar prosperity. And many of them were beginning to adopt a hedonistic subculture that dovetailed with the angst-ridden counterculture of the 1950s and 1960s. "Feelings of normalessness and apocalyptic fears pervaded Western youth in the mid-1960s," one historian notes. "These feelings and fears contributed to the reorientation of surfing as a revolutionary opt-out lifestyle."[44]

Some members of this class read the Port Huron Statement and joined Students for a Democratic Society; others took the less proactive path and alleviated their anxieties on the ocean's waves. There was something about the cultural style of California surfing that was especially suited to the task. Opposed to the attempt to control and conquer the wave, Hawaiian and Californian surfers generally embraced a Polynesian philosophy that fostered "a communion with nature." This anticompetitive style of surfing came to be know as "soul surfing" in the 1960s. Surfing, according to one big-wave rider, became "the ultimate liberating factor on the planet. You're working with nature in the raw in surfing."[45] The purpose and philosophy of the *Kon-Tiki* expedition participated in the same spirit as California soul surfers.

These three strands of Tiki culture—Polynesian camp, middle-class Pacific vacations, and surfing—tangled together on the heaving pubescent chest of *The Brady Bunch*'s Greg Brady, played by Barry Williams, while shredding the waves of Oahu's North Shore. *The Brady Bunch,* a popular situation comedy in the late 1960s and early 1970s, occasionally provided a portrait of the middle-class vacation: Disneyland, the Grand Canyon, and, of course, Hawaii. The three 1972 Hawaii episodes are given narrative force by Greg's discovery of a small Tiki idol that threatens to poison the family's vacation. Polynesian artifacts, surfing culture, and a middle-class vacation at the Sheraton Waikiki, with cameos by Don Ho and Vincent Price (who portrays, tellingly, an estranged archeologist), freely circulate through the episode's setting and script. Despite the culturally illiterate nature of the Bradys' experiences in Hawaii, the episodes perfectly capture the antitechnological, primitive, and anxiety-ridden interaction of white middle-class America with the ocean landscape.

What was the heart of Kon-Tiki fever? Why such a public reaction? And what about Trader Vic's, Hawaii vacations, and the cult of surfing? Heyerdahl's biographer suggests that it

> contains a strong element of the mystery of nature, the attraction of the sea, the urge toward the unfathomable, which the sea represents. What wonders shall we encounter if we venture out toward the horizon and beyond? What is hidden in the depths of the sea? Technological miracles like satellites and electronic brains naturally arouse our admiration, but they find no place in our emotions. The desire to return to a primitive life will always exist in most people.[46]

Another commentator pondered adventure stories like Heyerdahl's and suggested that the public "has wanted inspiration [that] diverts us from office and family problems to the problems of a man on [a raft]."[47] Tiki culture provided an oceanic balm for America's cold war anxiety. The desire to escape civilization for a wilderness experience was in no way new to the postwar era; but what was new was that Americans were increasingly turning to ocean seascapes for their wilderness experience.

FROM OCEAN HEALING TO HEALING THE OCEAN

The *Kon-Tiki* episode was a crucial moment in Heyerdahl's career, but it was only one of a number of such moments. He was unwaveringly devoted to his theory and continued to bolster support with research on Easter Island and various South American archeological projects. More generally, he widened his migration interests to consider the possibility that the world's oceans had long provided a highway for the meeting of various peoples throughout prehistoric times. The *Ra* expeditions (1969 and 1970) found Heyerdahl and new crews on ships constructed of papyrus reeds sailing from Africa to America. And amid the growing struggles of a world caught in the icy battle of two superpowers, he became something of a diplomat for calling attention to those forces that joined humanity. Sometimes he could be outright political, as when he burned the *Ra* as a ceremonial ritual intended to convince world diplomats that international cooperation instead of bloc politics was the order of the day. In 1966 he became vice president of the Association of World Federalists, an international organization that was incorporated in 1946 to help promote peace through world federalism.

Almost as an afterthought, but more likely a consequence of his global journeys, Heyerdahl called attention to the environmental problems that plagued the ocean. In the same way that he was able to observe the profusion of Pacific life on the *Kon-Tiki,* so too did he witness a daily spectacle of congealed oil and plastic trash on the surface of the Atlantic on the *Ra* and the *Ra II.* At the conclusion of the popular narrative of the latter expedition, he began to wonder how long

> whale and fish [would] gambol out there? Would man at the eleventh hour learn to dispose of his modern garbage, would he abandon his war against nature? Would future generations restore early man's respect and veneration for the sea and the earth, humbly worshipped by the Inca as *Mama-Cocha* and *Mama-Alpa,* "Mother-Sea" and "Mother-Earth"? If not, it will be of little use to struggle for peace among nations, and still less to wage war, on board our little space craft. The ocean is not endless.[48]

At the 1972 United Nations Conference on the Human Environment in Stockholm, Heyerdahl was invited to be an informal speaker on the subject of the vulnerability of the ocean. More than just a "big hole in the ground filled with undrinkable salt water," he began, we have come to think of the ocean as "a conveyer for ships, a source of food, a holiday playground and . . . it acts as a filter receiving dirty river water and returning it to our fields through evaporation, clouds, and rain." Moreover, it must not be forgotten, he insisted, that all life got its start from the ocean, and it continues to play a vital role in the meteorological and biological maintenance of the globe. Evidencing the increasingly sophisticated understanding of nature, he noted that "more than half of the biological pyramid supporting man at its apex is composed of creatures living in the sea. Remove them and the pyramid collapses; there will be no foundation to maintain life on land or in the air." The issue was given even greater moral force by the fact that the human onslaught of terrestrial resources will require a healthy ocean. "A dead sea means a dead planet."[49]

Heyerdahl was not the first to do so, but he was powerfully elaborating on Carson's oceancentrism. It is not a surprise to learn that Heyerdahl believed that, yes, the oceans were vulnerable; the oceans were used as a site for disposal of trash, nuclear waste, toxic chemicals, and sewage; overfishing and oil pollution also threatened the waters. Modern science and technologically advanced industrialism threatened to contaminate an ocean that had previously been considered a source of unlimited resources and a disposal sink with infinite diluting potential. He concluded with a note befitting of the international nature of the meeting. He called on delegates to "put aside all immediate personal and national interests" to deal with the problem: "Nations can divide the land, but the revolving ocean, indispensable and yet vulnerable, will forever remain a common human heritage."[50] Heyerdahl was always searching for an alternative amid the context of crisis and calamity.

JACQUES COUSTEAU AND TECHNOLOGICAL UTOPIAS OF THE OCEAN

Jacques Cousteau was searching for alternatives as well. Instead of balsa, reeds, and history, however, he entered the ocean with an array of technological apparatus that appeared extraordinarily modern, even futuristic. There is no way to easily take stock of a man whose life was so very apparent, though artfully constructed, to the world. Obituaries from his passing routinely highlighted the twin vectors of his life: popular explorer of the ocean, and ardent, even pioneering, environmentalist. He is often given credit for making us aware of the environmental stresses on the ocean levied by urbanization and industrialization. Although this is certainly true, there is another way of thinking about Cousteau's work that may help to explain his overwhelming popularity in American culture. The *Calypso* captain was less an exception and more a typical representative of another way American culture was taking stock in the ocean during the latter half of the twentieth century. Cousteau was the embodiment of a modern technology of freedom. His work and his representations often had less to do with fish and coral reefs than with the technologies that enabled the human movement into, and conquest of, the ocean frontier.

For sake of comparison, Cousteau had much in common with William Beebe. More than their similarly tall, gaunt frames and balding heads, they were both showmen who put themselves, their work, and the ocean into popular discourse.

Cousteau, however, never called himself a marine biologist, a scientist, or even a naturalist. Even his obituaries have failed to shed critical light on Cousteau's self-designation as an "oceanographic technician."[51] He is routinely touted as an environmentalist; some even call him the father of environmentalism. He had clearly outlined the dangers that modern existence posed for the ocean and for life in general, especially after 1970 when his work dovetailed with the vogue of modern environmentalism. But even here he was a technician, proposing technological schemes—some might use the term *appropriate technology*—that would save the ocean from humanity. Like Heyerdahl, Cousteau reflected an American culture that was grappling with the technological occupation of the postwar ocean; Cousteau just did it in a different way.

In the late 1930s and early 1940s, Cousteau could be found in the waters of the Mediterranean with a homemade diving mask, flippers, and speargun; indeed, coastal France was ahead of America's spearfishing craze by about fifteen years. Cousteau turned to swimming after an automobile accident that crippled his left arm. His love for swimming and diving grew, and the logical task at hand in the mind of Cousteau in the early 1940s was to extend the amount of time the human body could stay underwater. He had bonded with two diving enthusiasts, Philippe Tailliez and Frederic Dumas, and the three became a kind of informal research team with the goal of developing technology for extended underwater sojourns. Their efforts bore only mildly palatable fruit, until Cousteau's late 1942 collaboration with Emile Gagnan, an engineer with Air Liquide Cie., a French commercial gas firm.

Cousteau, like John Muir, seems to have had a childhood disposition for mechanics and engineering: childhood erector sets, a prototype of a battery-powered car, and, later in adolescence, motion cameras. But these were merely youthful, though important, pastimes, and the key to Cousteau's regulator problem fell to Gagnan, who had developed a demand valve that fed natural gas to automobile engines. The Bakelite mechanism, adapted to deliver compressed air into a diver's lungs, automatically adjusted the requisite air to compensate for a diver's trip through a variably pressured liquid environment. The Cousteau-Gagnan regulator was the centerpiece of the Aqua-Lung, which, according to Cousteau's first of many articles in *National Geographic Magazine,* "frees a man to glide, unhurried and unharmed, fathoms deep beneath the sea. It permits him to skim face down through the water, roll over, or loll on his side, propelled along by flippered feet."[52] Cousteau's Aqua-Lung was the first of a series of technological devices that extended this freedom for exploring, according to Tailliez, "the last frontier on earth."[53]

Cousteau asked for, and received, permission from the French navy to film his team's maneuvers with a makeshift underwater motion picture camera sealed in a watertight container. The resulting shorts, though of marginal quality, began to attract modest public interest. Through the publicity work of Perry Miller, then an American worker for the United Nations, *Life* magazine ran a short spread of grainy black-and-white pictures of the Undersea Research Group waltzing with an octopus, fending off a white-tipped reef shark, and exploring the underwater postwar wreckage of Toulon Harbor. There was a photo of a released underwater mine, and another of the eighty-pound "Cousteau camera."[54] The article had all the seeds of Cousteau's American popularity. Shortly before the 1950 magazine article appeared, Universal Pictures bought the rights to four Cousteau films, and thus he made his humble entrance into American culture.

The Aqua-Lung was also emerging in civilian sectors of western culture as well. In 1943 Liquid Air created a subsidiary known as U.S. Divers Inc., and with Cousteau as lifetime chairman of the board, the company became the premier distributor of scuba and affiliated diving equipment. No doubt it was a combination of entrepreneurship and an indefatigable love for ocean exploration that set Cousteau to the business of demonstrating the Aqua-Lung's value in a variety of fields. Like Beebe scampering around for a legitimate use for the bathysphere, so too did Cousteau pitch for converts to participate in an underwater world filled with cyborg "menfish." The centerpiece of the operation was a new oceanographic vessel, perhaps the world's most famous ship: the *Calypso.*

THE *CALYPSO*

The *Calypso* was the product of Cousteau's desire to "see our invention turned to more peaceful and productive use." To this end he required a "floating laboratory, workshop, and diving platform" that enabled the Undersea Research Group to extend its reach far beyond Toulon Harbor. With British and French patronage, he purchased a furrowed American mine sweeper, stripped it down to the hull, and refitted it with sonar apparatus, an aluminum flying bridge, a diving station off the stern, a false prow with windows, and an interior diving well for easy access to the ocean. The *Calypso*'s maiden expedition began on November 24, 1951. The ship's manifest listed a few members of the URG, a small crew for the *Calypso,* and at least five French naturalists. Their goal was the Red Sea coral island of Abu Latt off Saudi Arabia. The naturalists went about the business of observing the diverse flora and fauna of the tropical sea. The expedition's success laid not so much with the equipment's scientific value, but rather with its ability to capture on film the colors of the undersea world. The entire expedition made a stunning visual impression in *National Geographic Magazine*.[55]

A similar display of technological innovation and scientific research came together when the crew of the *Calypso* salvaged a Greco-Roman argosy in 1951 and 1952. The third-century B.C.E. ship was transporting wine-filled amphorae and pottery when it went down off the coast of Marseille. Cousteau collected monies from the French navy, the Ministry of National Education, Marseilles's political and financial elite, and the National Geographic Society. One objective was to salvage the vessel's cargo, raise the hull, and then, perhaps with a nod to Heyderdahl, "duplicate her exactly and sail her on her last course, from Delos to the scene of her foundering, using only replicas of her original fittings, rigging and navigational instruments."[56] The vessel proved far too fragile for such a plan, but the crew and divers of the *Calypso* spent a year pioneering the nascent field of underwater archeology.

The centerpiece of the operation was a suction pipe used for excavation. Cousteau noted that "sometimes we got astride it and felt it vibrate like a spirited horse's neck. . . . And, like a browsing horse, the mouth went forward into the pasture, munching shells, sand, shards, and things too big for it to eat."[57] The material traveled up through the suction pipe and into a filtering basket, around which a team of archeologists sifted through the detritus for antiquities and the occasional modern French coin (the *Calypso* divers became adept at practical jokes). Cousteau also extended the vision of the shipboard archeologists through the first effective use of closed-circuit television. The expedition made use of Harold Edgerton's strobe-flash cameras. Indeed, the collaboration between Cousteau and Edgerton—Gilbert Grosvenor brought them together during Cousteau's 1953 visit to Washington, D.C.—revolutionized underwater photography. Again, the *National Geographic* article was as much a feature of Cousteau's technological innovation as it was about a two thousand two hundred year old merchant ship.[58]

While in dry dock at Marseille, Cousteau was approached by a representative from the British Petroleum Company, who proposed an expedition to search for oil-bearing structures in the Persian Gulf. It was a lucrative and timely deal that allowed Cousteau to purchase a fishing boat that continued the archeological dig as *Calypso* fired up its inboards for a trip along the Red Sea, thence to the Persian Gulf, south to Madagascar, and then home. Again, the *Calypso* was outfitted with new equipment "to determine whether diving could contribute to submarine prospecting."[59] The destination was a concession area south east of the Qatar peninsula controlled at the time by Sheikh Shakbut Bin, sultan of Abu Dhabi.[60] The Cousteau team's dalliance with oil prospecting was brief but effective. Over the next two decades, residents of the

United Arab Emirates would reap a fortune in oil extraction, thus demonstrating the value of Cousteau's underwater technologies in fueling the energy desires of the developed world.

While he was busy finding new and inventive uses for his technologies, Cousteau continued to photograph and take motion pictures of the "virgin landscape of the undersea."[61] *The Silent World* was released in 1956, won the best picture award at the Cannes Film Festival, took an Oscar for best documentary, and was without question one of the most beautifully filmed portraits of the world under the ocean surface. With an ominous music score as background, the beginning of the film showed a bevy of divers, all conspicuously holding sulfur flares, headed straight down into the inky darkness of the sea. If Cousteau intended to give the impression that human exploration of the ocean was entering a new phase, he couldn't have had a greater effect. The brief underwater shots of Disney's *20,000 Leagues under the Sea* (1954) appeared paltry in contrast. Indeed, the helmet-diving, heavy-footed explorers of the cinematic version of the Verne story mark the distinction between the long history of clumsy human diving and the new era of effortless grace. The point is reinforced in the *Silent World* as one of Cousteau's crew approaches a Greek helmet diver busy making a living from the sea. The Greek diver returns to the surface and after being debriefed by the *Calypso* team, straps one of the Aqua-Lungs to his back for a try. The technological exchange underscores one of the most important, and overwhelmingly overlooked, aspects of the film and Cousteau's legacy.

The late 1950s and early 1960s were triumphant days of fame and technological prowess for the Cousteau team. They constructed a deep-sea submersible that brought a crew of two to the depths of one thousand feet in a kind of underwater flying saucer that looked like "a huge mechanical turtle with two gaping eyes. It has a steel hull and a top hatch. . . . A small port is provided for a motion picture camera. There is also a hydraulic claw to collect samples of marine life."[62] To play host to the saucer, in 1961 Cousteau christened *Amphitrite,* a Zodiac-built vessel composed of a nylon and neoprene sandwich, with new forms of fiberglass, foam plastic that was lighter than wood, and a new stainless tubing of aluminum and magnesium alloy.[63] These extraordinary technological devices set the stage for Cousteau's experiments in saturation diving, his most grand technoutopian project.

CONSHELF

Speaking to the members of the World Congress on Underwater Activities in London in 1963, Cousteau boldly announced that humanity was evolving, or

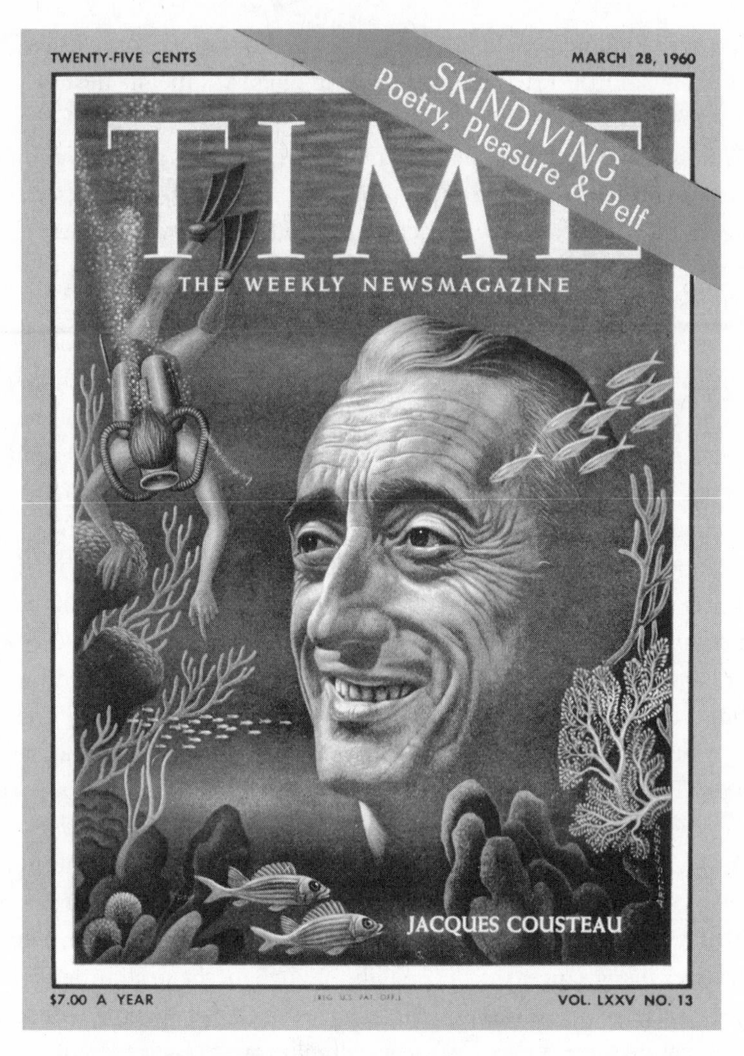

A 1960 cover of *Time* magazine showing Jacques Cousteau. Cousteau was at the leading edge of popular underwater exploration. Courtesy of Time Inc.

had the ability to evolve, into a new organism that he called *Homo aquaticus.* This was a man who could live and breathe underwater—a man, according to James Dugan, who "would dwell among his kind in submerged towns and swim about on his daily labors in the open depths." Such an evolutionary jump, or reversal, depending on one's perspective, needed to be helped along by technological know-how. This kind of "surgical transformation" was spurred, in Cousteau's words, "by human intelligence rather than the slow blind natural adaptation of species." He noted that NASA was constructing an "artificial gill" that would allow the oxygenation of an astronaut's blood without the pesky nuisance of breathing. The gill, "fitted under an arm and linked with the aorta by surgical manipulation," could easily be adapted for underwater exploration, though given the pressure of ocean depths, it would be necessary to pack "the rib cage with a non-compressible, sterile plastic." Cousteau noted that the incipient cyborg evolution was still about fifty years away, but the potential payoff would spur capital outlays and further research: "Virtually all the food and raw materials we could exploit on the ocean floor area exist in this zone of the continental shelf and the continental slope." With a nod to postwar Malthusian fears of overpopulation, he added that "we must create an underwater population to help accommodate our preposterous birth rate." *Homo aquaticus* would also begin the project of domesticating marine life as "ranch animals," a project that would transform fishing into "stock farming." But more than the economic potential, the evolution of *H. aquaticus* signaled "a life full of daily inspiration like that to which man has risen as a result of creative developments in his past—the Greek concept of the ethos, the High Renaissance, the 18th century revolutions."[64] A lot was at stake here, including the solution to a world caught in the banal and demoralizing climate of cold war anxiety. Though there were a number of doubters at the conference, dismissing his ideas as Jules Verne–like fiction, Cousteau's vision seems to have captivated many.

His announcement was actually a work in progress. The previous autumn, Cousteau had sent down two men to reside in a seventeen-foot cylinder filled with compressed air for a week. The thirty-two-ton structure was named *Diogenes* after the tub-dwelling Greek philosopher. It was lowered to thirty-three feet a few miles off the coast of Marseille. The divers freely moved from their temporary home into the water, where they donned Aqua-Lungs to work five hours a day. Conshelf I proved that humans could live for relatively long periods of time at sea and, instead of simply relying on quick jaunts from surface to ocean deep, increased the time and ease with which humans could engage with the ocean. Cousteau's description waxed philosophic. His overall goal, he wrote in *Popular Mechanics,* was to "free man from the slavery of the surface, to invent ways and means of permitting him to escape from natural limitations . . . to move about, react and live within the sea and take possession of it." This was especially pertinent to taking full military and economic advantage of the continental shelves, thus giving the project its name. *Diogenes* was his first invention that returned men to the sea, "the element from which the beginnings of man emerged some millions of years ago."[65] But this was just a humble test for grander ambitions.

Conshelf II, the 1963 facility, at thirty-six feet beneath the surface of the Red Sea, was brilliantly chronicled in one of the longest articles *National Geographic* ever published. There was an air of domestic bliss that perfectly overlapped with Eugenie Clark's attempt to make a home in the sea. The Starfish House was decked out with all the conveniences of modern life: air conditioning, closed-circuit television, and telephones.[66] Commentators put Conshelf II into an American context; for instance, a *Time* reporter noted that "Cousteau looks on the sea the way Daniel Boone looked on Kentucky, as a fine place to colonize. He thinks humans should do what porpoises, seals and other mammals have done, already: adapt themselves to underwater living and beat the conservative fish at their own game."[67] But to be sure, this was serious business. "It must lead to scientific research," Cousteau instructed his American audience, "to prospecting for wealth, and to greater utilization of the oceans. Finally, it must lead to human occupation of the sea floor." Cousteau did not hold on to illusions that colonization of the continental shelf was necessary to directly solve the world's population problem. The undersea world, in Cousteau's estimation, would become—indeed, was becoming—a new environment for human labor. These workers would manage the utilization of the continental shelf as a frontier region: oil, natural gas, liquid sulfur, gem diamonds, manganese nodules, and undersea fish farms, "analogous to stock ranches on land."[68]

The five oceanauts, as Cousteau was becoming fond of calling them, of Conshelf II wore distinctively reflective diving suits made of aluminized nylon-neoprene. When not scrubbing algae growth from the outside walls of the units, the oceanauts, like ranch hands, busied themselves with the capture of marine organisms and secured them in plastic pens for biological study. For the most part, the fundamental job of the deep-water crew was to be seen, and to be poked and prodded. Every day a physician dove to the Starfish House to take routine vitals. As with all of Cousteau's ventures, Conshelf II was massively photographed. Indeed, one of the primary functions of the diving saucer, other than a few routine maintenance tasks, was to film the village and the oceanauts at work. The motion pictures were dazzling and provided enough footage for Cousteau's second Oscar-winning documentary, *World without Sun*, originally titled *Conquest of the Sea*.[69]

A key outpost of the underwater village was Deep Cabin, a much smaller split-level unit that was the home of two oceanauts at a depth of ninety feet. If the Starfish House was the site of the main experiment, Deep Cabin was the pilot project for Conshelf III. This 1965 experimental site was deeper, far more dangerous, and far more practically oriented. The oceanauts were given specific tasks to perform; they were undersea oil riggers. Petroleum companies had watched Cousteau for years, going back to his early 1950s' experiments with the gravimeter. The French national petroleum office had even financed a large share of Conshelf II, as they did with the third experiment. But prospectors were going to get more for their money with the third experiment. Along with the innovative living quarters, a model of an oil wellhead was lowered to three hundred fifty-five feet below the Mediterranean Sea surface.

In mid-September 1965, the crew of six was sealed in the "wilderness outpost" while still at surface level. Internal pressure was increased to a depth of three hundred fifty-five feet, and then a dozen sea vessels and more than one hundred and fifty technicians oversaw the complex task of lowering the outpost to the seafloor along guide wires. This time the oceanauts were little more than test subjects: physiology, brain waves, endurance, sociability, and various other behavioral and biological data were accumulated by the team above. Simply put, they were engaged in an experiment to discover the limits of *Homo aquaticus;* could humans work for long periods of time at three hundred fifty-five feet without ill effect?

Cousteau boldly summed up the achievements of Conshelf III: "The oceanauts . . . had subdued a hostile territory" and had "helped to double the habitable and exploitable region of the world's continental shelf."[70] The boosterism masked an experiment riddled with problems; in Cousteau's estimation, the oceanauts solved more engineering problems during their undersea tenure than the crew of the *Calypso* had encountered since 1950. Poor weather stressed and threatened equipment on the surface. On day eighteen, a rebreathing pump exploded. Phillipe Cousteau—chief photographer, one of the oceanauts, and the elder's beloved protégé—dealt with constant lighting, turbidity, and dampness problems that hampered the photographic record.[71] Cousteau summarized the moral of the experiment, perhaps a little chastened when compared with his 1962 projections: "Conshelf Three was an epic of triumphant men and failing machines. In this day of automatamania, or worship of gadgets, the oceanauts served up a healthy reminder of how vastly superior to mechanisms old *Homo sapiens* remains."[72] This was new, and a far cry from *Homo aquaticus.* Cousteau came face to face with the stunning limitations of a technological universe.

He had originally planned two deeper Conshelf trials. It was likely the prohibitive costs and the technological problems that took the proposals off the table. But something else was stirring in Cousteau's heart and mind. He still had grand technologies on the drafting board—an artificial floating island that served as an oceanic monitoring station (the island was quickly lost to a fire), and a six-man diving bell designed for seabed oil prospecting.[73] But his personal adventure was about to take an exciting turn into the world of mass culture and modern environmentalism. Certainly technological optimism would continue to play a role in Cousteau's operations, but the spirit of that technology was subtly changing. Why? Cousteau and his team were not the only humans moving and living in the ocean with a technological sophistication.

American institutions, especially in the 1960s, had their own visions of the technological control of the world's oceans.

AMERICAN MACHINES IN THE OCEAN GARDEN

Cousteau's technological harnessing of the ocean was only the most visible human engagement with the ocean in American culture. His work as it appeared in magazines, books, movies, and television was a microcosm of some major changes in America's use of the oceans in the postwar period. Just as the Aqua-Lung continued to alter the way Americans played in the ocean, so too did scuba, along with a wide array of engineering wonders, enable the technological occupation of the ocean as an economic and military frontier. The oceans were bright with promise and possibility, a cache of resources in the water itself as well as under the seabed ripe for extraction that would more than supplement the consuming desires of an energy-intensive culture, and a geography of threat and freedom that would protect American cities from imminent nuclear disaster. You could follow a bit of the development by watching *Seawolf* and *Voyage to the Bottom of the Ocean* on television in the 1950s and 1960s.

A kind of utopian vision coalesced in the mid to late 1960s that brought the full weight of the military-industrial complex to bear on America's continental shelves and the wider deep sea. "The sea, man's first frontier," noted journalist William Smith in a 1968 *New York Times* article, "has become his last major earth-bound challenge. It has also become an important goal in the march for investment opportunity and profit." By 1966, it was estimated that the military was spending $4 billion on ocean-related preparedness; oil companies were investing $2 billion; another $3 billion was spent for marine recreation, and a modest $400 million was spent on commercial fishing. Americans also looked to the continental shelves for mineral resources to be mined to fuel economic growth. Other than military expenditures, the American move into the ocean was largely funded by private business at the same time that the federal government was spending large funds on the space program. The federal government's largest sea presence was military. Despite President Johnson's 1968 appeal that the Soviet Union join with the United States and all nations to exploit the ocean for all humankind, it didn't escape notice that the oceans, as space, were becoming a pivotal cold war geography. Smith implied as much when he quoted the statements of Petrovich of the Soviet Union: "the nation that first learns to live under the seas will control them. And the nation that controls the seas will control the world."[74] The issue became a paramount concern on Eisenhower's and Kennedy's watches.

One of the reasons for this optimism was American offshore drilling for oil and natural gas, a developmental imperative given the energy crisis of World War II. Ocean engineers were like the railroad designers of the nineteenth century, and federal and state governments continued to play their roles in settling the ocean frontier. Beginning in 1954, the federal government divvied up the continental shelves in quasi-homestead fashion (the word *seastead* made a brief appearance in the late 1960s).[75] Tracts of ocean geography were bid on by major oil companies; only the largest had sufficient capital to engage in such a speculative prospect. By 1968, the Department of the Interior had 1,276 leases. Monroe Spaght of the Royal Dutch/Shell Group called these areas "one of man's last terrestrial frontiers"; Chas Jones of Humble Oil and Refining Co. called the ocean depths a "mare incognitum."[76]

The U.S. navy, too, had long sought control of the seas. But after World War II the navy invested heavily in subsurface warfare. Vast sums of money of were earmarked for the development of nuclear powered submarines and submarines that could launch nuclear missiles. Perhaps the most popular manifestation of America's technological invasion into the ocean was the navy's Man in the Sea program that sponsored the Sealab experiments. The program was the brainchild of navy physician George Bond, a pioneer of saturation diving, but it was tragedy that breathed life into the Sealab program. On April 9, 1963, the nuclear attack submarine U.S.S. *Thresher* went down off the Newfoundland coast with one hundred nineteen men. America would have to find a way to rescue future submariners, to say nothing of salvaging precious cold war technology. The navy convened the Deep Sea System Research Group that, after eight months of investigation, proposed a five-year program that would tackle the problems of deep-sea salvage of objects, the recovery of submarine personnel, and the future habitability of the ocean floor. Late in 1963 undersea rescue and salvage technology became a navy priority. The Office of Naval Research approved Bond's Sealab proposal as part of the new Man in the Sea initiative. Plans for the habitat were drawn up at the U.S. navy's Mine Defense Laboratory in Panama City. By the following spring, Bond had assembled his team for the first trial close to Argus Island off the Bermuda coast.

Late in May, Sealab came to a gentle rest some one hundred ninety-two feet beneath the surface of the ocean. Bond put together a team of four aquanauts who were, in his estimation, "a breed apart. . . . They are the ones who take the chances for all humanity, who care nothing for fame but seek only the satisfaction of battling odds and attacking new frontiers."[77] Despite Bond's modesty, he was well aware that Sealab needed to be a popular event similar to the nascent space program, which was receiving massive federal funding. Throughout the 1960s ocean explorers found themselves in a kind of battle with space explorers, and the latter had a decisive advantage. Bond played the game brilliantly and saw to the participation of Scott Carpenter, the astronaut who piloted Aurora 7 of the Mercury program for three orbits around the Earth in 1962. At one point, Bond called Carpenter his "astronaut shield of protection," by which he seems to have meant a heroic and familiar face for public relations.[78] Carpenter had an untimely motorcycle accident and was scrubbed from Sealab I, though he would play an instrumental role in the success of Sealab II.

Like America's space program, the Sealab experiments were observed by the American press with similar enthusiasm. Reporters, photographers, and camera operators jostled among the technicians and engineers to record and publicize the heroic missions. Press conferences and television interviews consistently dogged Bond and other leaders of the project. There were some special flourishes to the daily routine of the trial. Tuffy, a porpoise trained by navy personnel, acted as a courier between the surface and the facility, one of the more benign uses of marine life by the U.S. navy. Carpenter also had dramatic conversations with other frontier explorers. He spoke with several of the oceanauts in Cousteau's Conshelf III, which coincided with Sealab II; and perhaps most stunningly, he exchanged greetings with Col. L. Gordon Cooper Jr., who was in Gemini 5 passing a hundred miles above the earth.[79]

For the most part, the true objective of Sealab was to observe the physiological and psychological stresses of humans working underwater. The work of the aquanauts themselves played second fiddle. One team performed experimental salvage operations and installed an underwater weather station. Another team included a number of oceanographers from Scripps Oceanographic Institution. After the trial, biologist Arthur Flechsig noted that the type of science conducted was fairly routine, "old-fashioned biology—like going on a nature walk." Captain L. B. Nelson, second in command, said that Sealab "will give man the abilities to develop new concepts that have not been and cannot be developed above the surface of the watery realm." However, it "will give mankind a technique for further expansion."[80] On March 18, 1965, shortly before Sealab II began its work, ABC presented a documentary entitled *Man Invades the Sea*. The report chronicled Sealab I and Cousteau's Conshelf programs, and gave a list of future projects. The program described the sea as one of the "last remaining storehouses of such natural resources as food, minerals and oil" and therefore helped to push along a wave of ocean engineering enthusiasm that was amplified by the stunning success of Sealab II.[81]

Sealab III was America's most extravagant push to inhabit the ocean. The $10 million project led to a fifty-seven-foot, three-hundred-forty-ton, bright yellow habitat that one reporter described as a railroad car without wheels.

Sealab II was lowered in the Pacific in 1965. It was nicknamed the "Tilton Hilton" and housed aquanauts from the U.S. Navy. Courtesy of NOAA Photo Archive.

Five teams were to spend twelve days apiece at a remarkable six hundred ten feet off San Clemente Island. Miami Seaquarium obtained a contract to train eight porpoises for the effort. Robert Cowen remarked for the *Christian Science Monitor* that "they are literally going out to live and work in earth's last great untamed wilderness."[82] The affair was scheduled for spring 1968 but was pushed back a year while the leadership team dealt with a barrage of technical complications. The problems continued as the team lowered the facility to the ocean floor: air leaks, flooding chambers, malfunctioning transports. On February 17 a team of four divers went down to fix some of the problems, and early in their work, Berry L. Cannon began to experience seizures. The other three divers brought him quickly to the surface decompression tank, but Cannon died on the way up. An investigation found that Cannon's breathing apparatus was missing the vital chemicals that scrubbed out carbon dioxide. There were even charges that a saboteur was wreaking havoc on the team. Just like Cousteau's Conshelf experiments, Sealab III came to a quick end, as did the Sealab program. Captain Nicholson thought that Cannon's death raised serious doubts about the human ability to function in deep waters for long periods of time.

The final great movement into the ocean in the 1960s was the widespread and enthusiastic development of submersibles. Although Cousteau's diving saucer certainly provided impetus for Americans to respond with submersibles of their own, it was more likely the mandate from the Deep Submergence Systems Review Group that provided the largest push. Since 1964, the DSSP was in the business of developing various deep-sea recovery vehicles (DSRV) like the *NR-1*, the *Trieste II*, and the eponymous class of submersibles called DSRVs (featured in *The Hunt for Red October*). Cousteau even served as a consultant to the project when visiting La Jolla in 1964. The same complex of companies that created America's fleet of nuclear-powered submarines also went into the submersible business in the mid to late 1960s. Occasionally, oceanographic institutes like Woods Hole worked hand in hand with the Office of Naval Research to develop submersibles like the famed *Alvin*.

From resource extraction to undersea warfare to the Man in the Sea initiative, oceanology—or ocean engineering—provided the language, concepts, and tools that shaped much of America's postwar relationship with the ocean. But this general air of ocean enthusiasm shouldn't be confused with consensus. Indeed, by the late 1960s a number of critical voices began to dampen the spirits of ocean pioneers. The litany of technologies gone wrong—the Texas Towers, the *Thresher*, Sealab, and Johnson Sea Link—did more than underscore that undersea exploration was a dangerous endeavor. The events spurred a discussion on the limitations of the human invasion of the ocean. President Johnson weighed in on the issue in 1968 when he addressed a Geneva audience and pleaded that the ocean floor be barred as a location for hiding nuclear missiles, just as it had already been agreed to ban weapons from space and Antarctica.[83]

Johnson's voice joined with civilian criticisms of oceanological optimism. *Wall Street Journal* writer Kenneth Slocum noted in 1968 that "despite considerable hoopla in recent years about the great underwater wealth presumably about to be tapped as technology enables men to conquer what some enthusiasts call 'the earth's last frontier,' there's little indication that the dream of harvesting vast quantities of food, drugs, minerals and petroleum from the oceans is anywhere near fruition." Robert Briggs, a vice president for an undersea mineral prospecting company, told Slocum, "Oceanography has been distorted all out of proportion as an economic event." Edward Wardwell, vice president of Oceans Systems Inc., sold all of his personal stock in ocean industries and noted that "people who think it's a get-rich-quick operation are in for a rude awakening."[84] The physical limitations of the human body in deep water were becoming painfully obvious. The prohibitive capital required for deep-sea economic ventures began to scare off companies hoping for a quick payoff. Just like Cousteau, American ocean explorers were coming up against the stunning limitations of human ingenuity and technology. *Homo aquaticus* was slow to evolve. But there were other reasons to sound a note of caution in the late 1960s; the human touch on the ocean environment didn't prove itself as benign as previously expected. The change in attitude occurred at precisely the same time that Cousteau was transformed into a legend.

THE UNDERSEA WORLD OF JACQUES COUSTEAU: REQUIEM FOR THE OCEAN

On August 17, 1966, the day before networks began televising recent pictures taken by the Lunar Orbiter 1, the *New York Times* announced that ABC was financing twelve episodes that were to be titled *The Undersea World of Jacques Cousteau.* Bud Rifkin, vice president of a television production studio, championed the series and negotiated a deal with ABC for $4.2 million to produce twelve one-hour shows over the course of three years. The money went for a *Calypso* makeover and new exploratory toys, like the two new P-500 minisubs called sea fleas, and a hot air balloon used by Philippe for aerial footage. New Aqua-Lungs, new wetsuits, new scooters, new cameras and lights, new Zodiacs, and new red tukes that became the signature headgear of the *Calypso* crew—all were provided. From the start, there didn't appear to be much of a thematic vision for the show other than guided spectacle; Cousteau's unofficial biographer called them "nature adventures." When queried on his objective, Cousteau simply responded, "What I have to do—and know I can do—is to make our discoveries more intelligible."[85] Cousteau traveled with crew and cameras to both familiar and new ports of call; often he invited scientists, perhaps to give the resulting shows an air of authority.

The first program, which predictably profiled sharks, had a cameo of Eugenie Clark performing experiments in the open sea. For the most part, *Sharks* (1968) was familiar fare: scenes of the shark cage in action, a test of Shark Chaser (hoped to be chemical shark deterrent), a few tagging experiments. The program received a Nielsen rating of 24.2, not the highest documentary rating of the year (which went to Jules Power's documentary *How Life Begins*), but large enough to attract the continued support of sponsor B. F. Goodrich and eventually Du Pont. Reviews of the program were less than glowing; one noted that Cousteau struck the pose of an "oceanic showman" who "contrived to inject a certain amount of drama into the presentation."[86] If he was ever bothered by such criticisms, he may have found some solace in Beebe's experience as a popular explorer.

Like *Sharks,* the Cousteau programs through 1968 and 1969 lacked the environmental sensibility that would be the Cousteau hallmark for the rest of his career. The first six programs, which ranged from the episodes on sharks, a feature on coral reef life, two on searches for lost treasure—one was in Lake

Titicaca—and a playful exposé on the natural history of two charismatic fur seals (named Pepito and Cristobal) were constructed around the format of "nature adventure." Even the episode on whales gave only a tip of the hat to their history of tragic encounters with the human species. Many of the programs focused on biological phenomena; shows were therefore organized around sharks, whales, manatees, sea lions, penguins, salmon, and a wide variety of other organisms. But there was a more noticeable format change that took place early in the program's history: Cousteau was going green. It appears in the late 1969 episode on "The Desert Whales," a feature on the gray whales that came close to extinction during the nineteenth century; the same whales that brought Roy Chapman Andrews to the waters off Japan and Korea.

The Undersea World of Jacques Cousteau became a kind of environmental travelogue, and throughout the 1970s Cousteau continued to chronicle the destructive effect humans were having on the ocean and on nature in general. The *Calypso,* a ship previously geared for research and general exploration, was fast becoming a mechanism for taking the pulse of the oceans. The litany of human misbehavior has an all too familiar ring to the modern ear: overpopulation, industrial and agricultural pollution, wastefulness, thoughtlessness, and arrogance. Cousteau's shows touched on all these. For those of us who grew up watching the reruns on cable TV, this was our first introduction to modern environmentalism. The compelling thing about Cousteau, however, was that he provided answers—or at least he tried.

If industrialism, with science and technology as handmaidens, got humanity into this mess, Cousteau had equal faith that it would get us out. Even with an environmental focus, the program continued to highlight the sophisticated gadgetry of the Cousteau team. Indeed, Cousteau was often criticized by commentators for using the *Calypso* to mount "formidable marine safaris" with "ultrasophisticated equipment."[87] Cousteau's seemingly endless shots of crew, ship, apparatus, the new PBY (a military plane capable of water landings), the new helicopter, the Zodiacs, and the balloon—"instruments that would have flabbergasted Ahab"[88]—led some commentators to accuse him of "self-advertisement."[89] More recently, critics have charged the Cousteau team of environmental abuses and insensitivity, and the always-present accusation of "staging" natural history. Although the charges may be justified, they distract us from a more important point: Cousteau was delivering a message in these programs. Science and technology were among the tools that led to the ocean's environmental degradation, and, Cousteau implied, they would also play a role, when properly used, in healing them. The *Calypso* machine, with its charismatic crew playing with sophisticated technological devices, was giving the oceans a helping hand. Technologies of play were transformed into technologies of salvation.

The television series was a reflection of Cousteau's wider efforts at environmental reform in the realms of policy and diplomacy. Why 1970? Didn't Cousteau recognize ocean pollution before that time? Of course he did, and he would often comment throughout the 1950s and 1960s, without any real effort at critique, on the irony of Greek amphorae, for example, resting next to beer cans. One of Cousteau's biographers believes that the 1969 Apollo deep-field photographs of the earth—the big blue marble—made an indelible impression on Cousteau that provoked his consciousness. Cousteau dated the epiphany to 1968 on his second visit to the Indian Ocean's Assumption Island; he had first visited it some thirteen years earlier, so he was in a position to assess change over time. He observed the damage to the coral island and came to appreciate the "fragility of marine ecosystems."[90] It is just as likely that the April 22, 1970, the first Earth Day, and the associated legitimization of political environmentalism, presented Cousteau with a perfect theme that would focus and refine his oceanic explorations and representations. Whatever the cause, the message was clear and widespread.

In 1971 the U.S. Senate was conducting hearings to establish the marine equivalent of NASA, the National Oceanic and Atmospheric Administration. Cousteau offered testimony that "the oceans are dying." "We are facing the destruction of the ocean by pollution and by other causes," destruction of coral reefs, declining populations of fish, pollution by continental runoff of pesticides and other industrial waste products. "One may wonder," he continued, "why so little care has been given to the ocean. The reason is simple. People have thought that the legendary immensity of the ocean was such that man could do nothing against such a gigantic force."[91] Western industrialized nations, he argued, need to discontinue their faith in the infinite diluting capabilities of the ocean. Simultaneously, industrialized countries must disabuse themselves from the idea of the "endless riches of the sea." He remarked in a 1970 acceptance speech for the Howard N. Potts Medal by Philadelphia's Franklin Institute, "It is fashionable nowadays to talk about the endless riches of the sea. . . . The ocean is regarded as a sort of bargain basement, but I don't agree with that estimate. . . . This moisture is a blessed treasure, and it is our basic duty, if we don't want to commit suicide, to preserve it."[92] Cousteau was echoing Carson's oceancentrism. Indeed, the ocean dominates the Earth and the life on it, but in the 1960s and 1970s, explorers and scientists were coming to grips with the real possibility that western industrial powers possessed the ability to destroy even the mighty sea. Similar lessons would have to be relearned in the 1980s, when ozone layer deterioration and global warming became objects of environmental discourse.

Cousteau was not about to give up the technological wonders that he had spent thirty years creating and refining, and in this, he was part of a much wider landed tradition. Modern environmentalism and technological innovation were not necessarily mutually exclusive, though there were plenty of critics making this point beginning in the 1950s and 1960s. In the 1970s there were two dominant attitudes that brought environmentalism and technology together in a complementary relationship. The ecological technocrats believed that the environment was a vast machine that could be controlled through the manipulation of environmental variables. The turn of a number of physiognomic dials and a few flips of trophic switches would, they hoped, return the ailing environment to a balanced state. In this way, the environment became a machine, and the role of humans was to use technological tools to further engineer the environment.[93] There was this in Cousteau, though he sometimes railed against the arrogance of modern technocrats. But his work more closely resembles the attempt to define a world filled with "appropriate technology." This movement began in the 1960s and was best represented by the *Whole Earth Catalog,* an early attempt at green consumerism that promoted the use and development of benign technologies, especially computers, alternative energy, and natural fibers.[94]

Cousteau is sometimes portrayed by critics as an individual whose environmental ethic contradicted his own interaction with the ocean planet. It is more helpful to understand how technology joined these two competing stories into a continuous tradition. In contrast to Heyerdahl, who based his ocean environmentalism on a kind of return to primitivism, a turning back of the clock, Cousteau was an ardent progressive who believed that innovative science and technology would be the Earth's only opportunity for redemption, a point made with spectacular clarity in an article he wrote for the *Saturday Review World.*

By the time this 1974 article appeared, Cousteau's critique had developed into a diatribe against western civilization. He lambasted our "exhilarating climate of pride" that demanded more food, more goods, and more energy. Ignorance coupled with the quest for power, "a disease of the cancer family," had led directly to the malady of the oceans. If a surgical excision proves successful, Cousteau argued, then humans would be able to forge a new relationship with the ocean, but via a relationship no less technological. He prophesied the creation of a world "weather production center" that would coordinate satellite and ocean-based data with certain ameliorative manipulations. The plan sounded a little like science fiction, with, for example, rockets spreading a biodegradable dye over parts of the ocean that would then promote evaporation followed by a cloudburst over a famine-stricken patch of land. Such technological euphoria, however, was based on the presumption of far-reaching international cooperation in which the ten to fifteen richest industrialized countries, the greatest polluters, give up a portion of their sovereignty to create, as Heyerdahl no doubt would have approved, an ocean high authority that would regulate the use—and misuse—of the world's oceans.[95]

The fiery article was published just as Cousteau was finishing his plan for a new U.S.-based nonprofit organization dedicated to research and education for the "protection and improvement of life."[96] Like so many of the environmental groups mushrooming over the American landscape in the 1970s, the Cousteau Society brought together a sizable constituency in an effort dedicated to education and political action. Under the capable leadership of Philippe, who had a strong influence on his father's green turn, the society's objective was to "bring the condition of the 'water planet' to public attention, while illuminating what is sublime, what is mysterious, what is crucial to our well-being but hidden from most of us."[97] Proceeds from membership dues went to finance the *Calypso*'s new environmental mission so as to break Cousteau's previous commercial connections. The society reproduced films and other educational materials for distribution to America's grade schools. Starting in 1976 Cousteau publicly bolstered the society by holding "involvement days," public teach-ins modeled after Earth Day, intended to educate America's youth. The society's literature also took a turn toward political action. For instance, the first edition of *Calypso Log*, the first society publication, criticized the movement at the 1974 U.N. Conference on the Law of the Sea to support exclusive economic zones that would lead to the sea being "recklessly parceled and dredged for profit."[98] The society urged that the United States back the idea of the ocean as a common responsibility for all nations.

This technological vision of environmental reform was put on brilliant display in the society's six-episode television series *Oasis in Space*, partially funded by the Public Broadcasting System. It was a show that brought the Cousteau Society together with NASA, civilian scientists, government agencies, and industry in order to tell a story about "earth and space, oceans forests, life; and it is about the incredible technology that is guiding us into the future."[99] The program had a radical edge unfamiliar to viewers of *The Undersea World of Jacques Cousteau:* industrial effluent, toxic waste, starvation, population explosion. Here Cousteau was incorporating the human side of the globe's environmental woes. The programs had a tremendous social focus, reminiscent of Cousteau's advocacy for justice issues that pertained to lesser-developed countries. Episodes detailed the nature-damaging impacts of strip mining and taconite processing, but also the consequences of poverty and hunger in the world. If the show focused on the destructive effects of modern technology, the Cousteau team was searching for a new cooperative spirit, a humane and appropriate technological world.

Cousteau, a man never afraid to create neologisms, called this new spirit *ecotechnie*. Both ecology and economics, he declared, "have the same duty: the art of harmoniously managing our household, the water planet earth. Both can do little without the help of technology."[100] The concept was partly born out of a special policy committee of the Cousteau Society's Council of Advisors, convened in 1980 and chaired by Dr. Elie Shneour. Ecotechnie was a global philosophy, a new ethical system, and a new social contract geared toward environmental sustainability and social justice for the preservation and health of future generations. The academic discipline of ecotechnie has never enjoyed much success in the United States, which is not surprising given America's general disinclination to participate in social and environmental reform on the global scale. Ecotechnie has, however, become a viable program of study at about a dozen European universities and is an affiliate of UNESCO. The new discipline of ecotechnie, along with its most popular symbol, the *Alcyone*—a partially wind-powered ship with high-tech turbosails—underscored Cousteau's progressive vision of an ocean controlled, monitored, and redeemed by modern technology.

Technology has been a central part of America's founding mythology at least as far back as the early nineteenth century.[101] Human ingenuity in the mechanical arts would win over the frontier and "improve" it for productive use. We begin to see a countermovement in the mid to late nineteenth century, a desire to cast aside human contrivances to embrace a purer, simpler, and more genuinely natural life. In reality, the two positions were never polar opposites, but as symbols, technology and the disparagement of technology have both worked to construct the meanings and practices of Americans encountering frontier regions. The ocean absorbed this legacy, especially in the postwar era. In their own unique ways, Cousteau and Heyerdahl became popular symbols of modes of thought and physical engagement that brought the ocean flooding into America's national imagination. But as history has shown, frontiers—and the people, plants, animals, water, air, and even topsoil in those frontiers—often pay a heavy price for national inclusion. Heyerdahl and Cousteau came to realize this in the 1970s. Their voices were part of a growing chorus of scientists, naturalists, politicians, and bureaucrats who were calling for a new ocean ethic.

Conclusion

In 1970 a team of five female aquanauts, all naturalists and scientists, were spending two weeks in yet another underwater habitat project called Tektite 2, a joint Department of the Interior–NASA affair. At fifty feet below the Caribbean's surface, Tektite 2 was a bit more humble than the expensive saturation diving experiments like the Conshelf and Sealab series—less bravado and more deliberate science. The fact that a team of five women was participating in an activity previously restricted to men drew some attention from the American public. *Life* ran a feature story entitled "Nest of Naiads," an allusion to Greek mythology's water nymphs. One photograph, taken inside the unit, shows the scientists gathered in the living area. The team engineer, Margaret Ann Lucas, is hanging an environmentally themed mobile; a cardboard representation of the earth crowns the display from which dangle a sickly fish, a drop of blood, and a declaration: "Pollution kills . . . Everything! Save our Environment. Fight Pollution Now!" The scientists were doing the routine work of marine biology, but they were also, in their own ways, environmentalists—students of and advocates for the natural environment. The team leader was Sylvia Earle, who noted in her first of many *National Geographic* publications that environmental degradation was a problem that affected everyone. The purpose of Tektite 2, in Earle's estimation, was to examine the ecological processes of Great Lameshur Bay's healthy marine environment to come up with a model for correcting "disturbed and unbalanced parts of the undersea environment."[1] For these aquanauts, this style of underwater exploration was part of the scientific side of modern environmentalism.

It helps to think of the Tektite projects as a key point of transition. The Sealab trials of the 1960s gave way to the scientific environmentalism then being embraced by many Americans. Indeed, by the first Earth Day on April 22, 1970, Carson, Clark, Heyerdahl, and Cousteau had already spoken, or were speaking, out against ocean pollution. Supreme Court Justice William O. Douglas, an ardent defender of the environment, had written an important article calling for the international regulation of ocean resource extraction and pollution. Wesley Marx's *The Frail Ocean* (1967) was by far the most comprehensive journalistic

All female crew-scientists aboard Tektite II, 1970. The navy and the U.S. Department of the Interior sponsored the saturation diving experiments to do research on people living in hostile environments. But it was also the beginning of scientific research influenced by environmentalism. Courtesy of NOAA Photo Archive.

treatment of the many human-induced threats to the ocean. Even the federal government was moving quickly forward on developing a national policy toward the ocean. The famous Stratton Commission published a report in 1969 that eventually led to some sizable restructuring in the executive branch as well as the creation of the National Oceanic and Atmospheric Administration (NOAA). Moreover, the report, *Our Nation and the Sea: A Plan for National Action,* contained what can only be called a restrained and cautious stance regarding the environmental pressures on marine environments. It remarked that many boosters throughout the 1950s and 1960s hailed the ocean as the "'last frontier' to be conquered by man." The commission called for modesty, thus reflecting a new concern "based on growing appreciation that the environment is being affected by man himself, in many cases adversely."[2] Several years later, groundbreaking legislation, such as the 1972 Marine Protection, Research and Sanctuaries Act, was signed into law. The environmental goals of the Tektite project fit squarely into this massive transition.

Why is it, then, that about forty years later, we are hearing that the ocean is in more trouble than ever? Overfishing, coral reef destruction, chemical pollution, the potential shifting of ocean currents, the creation of dead zones—all these are increasingly in the news. The Pew Oceans Commissions outlined the many dangers in its 2003 report, *America's Living Ocean: Charting the Course for Sea Change.* A year later, the U.S. Commission on Ocean Policy published its much anticipated *Ocean Blueprint for the 21st Century,* which called for a rededicated effort to coordinate ocean policy. The reports of commissions take on a more poetic voice in Sylvia Earle's plea for an ocean ethic, *Sea Change* (1995), which is echoed in Michael Berrill's *The Plundered Seas* (1997), Carl Safina's *Song for the Blue Ocean* (1998), Colin Woodard's *Ocean's End* (2000), Richard Ellis's *The Empty Ocean* (2003), and Callum Roberts's *The Unnatural History of the Sea* (2007). Reports on the loss of ocean catches and coral reef destruction regularly issue forth from scientific journals. Some studies have become a bit more prescriptive, some even optimistic—Rodney Fujita's *Heal the Ocean* (2003), John Field's *Oceans 2020* (2002), and Linda Glover's *Defying Ocean's End* (2004), just to name a few. Popular culture echoes this concern, coupled with climate change, in films such as *The Day after Tomorrow* (2004) and *Happy Feet* (2006). Even philosophers are getting into the game in order to develop rational systems of thought for extending a land ethic into the ocean.[3] Though the stakes might be higher, today's dire warnings and calls for reform feel peculiarly familiar. We had this conversation close to forty years ago. What happened?

Answers to this important question are better left to historians of ocean policy and international diplomacy. But, in all fairness, there have been some notable achievements. Some fisheries have recovered after cataclysmic crashes; a few charismatic species—like the green sea turtle and several species of whales—have also come back from the brink. Marine protected areas and sanctuaries also seem to be saving small amounts of fragile coastal ecosystems. Beaches tend to have less tar on them, and cans of tuna fish tend to be a little more friendly toward dolphins. By and large, these are only modest gains. Indeed, the story of the late twentieth-century ocean is a global narrative; the cast of characters is long, but it is clear that during the last forty years, America and the world as a whole have made steadily increasing demands on an ocean that has continued to be viewed as inviolable and unconquerable. These demands are both endorsed and challenged by the many constructions of the ocean in the American imagination.

This book outlines some of the ways the ocean has become a meaningful place in the American imagination. But it has not said as much about physical geographies. The ocean, just like the American West and outer space, is a real place.[4] A frontier or a wilderness is imagined space; they are ways to conceptualize actual space. The frontier is a mental framework that refers to real space but through a vital and necessary cultural lens. The primary purpose of this book has been to examine some of those lenses over the course

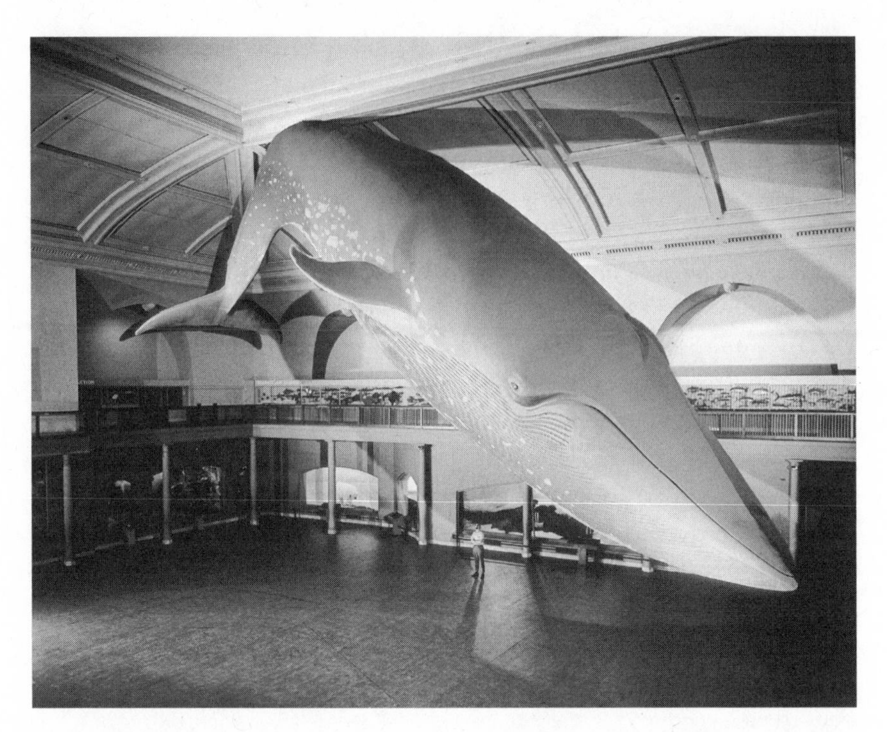

Blue whale in the Hall of Ocean Life in the American Museum of Natural History. In 1969 this model replaced Andrews's sulfur-bottom whale and signaled the dawning of a new era of environmental concern. Courtesy of the American Museum of Natural History.

of the twentieth century. Roy Chapman Andrews drew from his elite sensibilities to make a hunting ground of the ocean, and when not hunting, he was finding ways to harness the business tools of the day to more efficiently harvest whales from the ocean. Robert Cushman Murphy also wanted to conserve scarce natural resources, but he did so as a natural historian who cast a critical eye on the heedless changes that modernity was bringing to the ocean and ocean environments. In the process, he rationalized the ocean into a system of zones and borders; unintentionally, he made the ocean more legible for commercial interests. William Beebe found his way into the ocean by accident, but when he got there, he became the center of a publicity machine that brought the wonders and mysteries of ocean life to American audiences. These were not merely fish, Beebe would have us believe; they were exotic creatures from another world. There was a bit of this in Carson's brand of oceancentrism; she wrote in the context of World War II's aftermath and reminded America that the ocean was great and powerful and beyond our control—despite all of our efforts. Eugenie Clark created a different ocean—one that was warm and friendly, not a dangerous frontier, but an inviting wilderness for recreation fun and family bliss, a perfect container for the anxieties of the cold war. This, too, preoccupied Thor Heyerdahl, who reinforced Clark's domesticated ocean, though with simpler technology. While Heyerdahl's ocean beckoned America's middle class to find the rhythm of its waves, Cousteau's ocean called us to create a technological universe so that we could better understand, harness, and play in the ocean.

Most of us relied on pieces of their popular culture to come to know a little about this almost completely inaccessible geography. And these meanings became practice; they defined not only our attitudes, but also our behavior toward and treatment of the ocean. So in our current moment, as we stand witness to all too real ocean decay, what are we to make of our explorers? Were their voices ineffective? Yes and no. As I have shown, ocean explorers were unable to transcend the culture to which they belonged. And that culture—human, western, and American—was and is thoroughly rooted in terra firma. Sometimes explorers were endorsing the use of ocean resources that would eventually prove unsound; other times they were criticizing uses of the ocean that were profoundly unsustainable; and still other times they advocated for new attitudes and practices that would ameliorate the human touch on the ocean. But those endorsements, criticisms, and advocacies were part of a narrative already scripted through a history of westward expansion and Manifest Destiny. They literally saw and represented the ocean in a manner that history and their own culture demanded. It is in this light that our ocean explorers become part of a wider story—a replicated story of Americans using, working, and playing on the land.

Can the transformation of the ocean into a wilderness frontier truly be a reason for the failure of a benign human-ocean relationship to materialize in attitude as well as practice? To answer the question, let us take our cue from a drier history of wilderness. Up until the eighteenth century, wilderness frontiers were places to be feared, banished to, avoided, and conquered, or some combination thereof. European Romantics taught us that this wildness is not to be dreaded, but preserved and utilized for the regeneration of the western soul. In America this movement began with the Transcendentalists of the mid-nineteenth century, and by the late nineteenth century, this new notion of wilderness was codified into law. This was the beginning of the wilderness movement. It was an important moment, and according to historian William Cronon, our longing for distant wilderness spaces served as a distraction from the environmental promises and problems of the natural world in our literal backyard. In short, in our attempts to reinvent the wilderness as a place of moral value, we were actually "getting back to the wrong nature."[5] We may have been better off tending to the gardens in our backyards instead of praying to mountain cathedrals in the hope of spiritual revitalization. The argument casts a shadow on the history of ocean exploration as well.

As I have argued, ocean explorers and a great many Americans turned the ocean into a wilderness frontier—a geography ripe for exploration, exploitation, and, inevitably, conservation and preservation. All of these ideas are central to the history of America's experience with wilderness geographies. And Americans were adept in creating new frontiers in the twentieth century. Herein is the problem. In our zeal to transform the ocean into a wilderness frontier, we rarely wondered whether we should be engaged in the project in the first place. Ocean explorers, like all explorers in the western tradition, play a pivotal role in helping to create those places that call for mapping, exploitation, admiration, veneration, recreation, conservation, protection, and preservation. Perhaps one day we will be able to explore alien environments like the ocean without our terrestrial predilections, without the need to drain the ocean and pour it into the imaginary framework of a frontier wilderness. Until then, we can only view Lord Byron's stanza as either naive or overly optimistic. Actually, "Man marks the earth with ruin," but his ruin did not stop at the shore. He just looked to another new horizon to find another new frontier.

Notes

INTRODUCTION

1 Sylvia Earle, "Atlas of the Ocean," Radio Expedition sponsored by National Public Radio and the National Geographic Society, http://www.npr.org/programs/re/archivesdate/2001/nov/20011127.gulfmex.html, December 24, 2002.

2 Roger Rufe, "Something Old, Something New . . . Always Something Blue," *Blue Planet Quarterly,* summer 2001, 4. The most contemporary statement on the link between the ocean and the frontier—that is similarly imbued with an environmentalist message—is David Helvarg's *Blue Frontier: Saving America's Living Seas* (New York: W. H. Freeman, 2001), but also see the work of Sylvia Earle, Carl Safina, and Colin Woodward.

3 *Washington Post,* April 22, 2002.

4 Peter Neill, ed., *American Sea Writing: A Literary Anthology* (New York: Library of America, 2000), 58, 119, 200.

5 Nineteenth-century ocean exploration did, of course, occur. The United States Exploring Expedition, the Coast Survey, and the Bureau of Fisheries were notable examples. And although each of these enjoyed attention in the public eye, it's instructive to remember that Herman Melville's *Moby-Dick* was not recognized as an American classic until the twentieth century.

6 Theodore Roosevelt, "Conservation," in *The New Nationalism,* by Theodore Roosevelt, ed. W. E. Leuchtenburgh (New York: Prentice-Hall, 1961), 49–76.

7 David M. Wrobel, *The End of American Exceptionalism: Frontier Anxiety from the Old West to the New Deal* (Lawrence: University Press of Kansas, 1993).

8 Kerwin Lee Klein, "Reclaiming the 'F' Word, or Being and Becoming a Postwestern," *Pacific Historical Review* 65 (May 1996): 179–215; Stephen Aron, "Lessons in Conquest: Towards a Greater Western History," *Pacific Historical Review* 63 (May 1994): 125–48; David Wrobel, "Beyond the Frontier-Region Dichotomy," *Pacific Historical Review* 65 (August 1996): 401–29; Michael Steiner, "From Frontier to Region: FJT and the New Western History," *Pacific Historical Review* 64 (November 1995): 479–503.

CHAPTER 1: THE OCEANIC HUNTING GROUNDS OF ROY CHAPMAN ANDREWS

1 Roy Chapman Andrews, *Under a Lucky Star: A Lifetime of Adventure* (New York: Viking Press, 1944), 90–91.

2 Charles Gallenkamp, *Dragon Hunter: Roy Chapman Andrews and the Central Asiatic Expeditions* (New York: Viking, 1991); Ronald Rainger, *An Agenda for Antiquity: Henry Fairfield Osborn and Vertebrate Paleontology at the American Museum of Natural History, 1890–1935* (Tuscaloosa: University of Alabama Press, 1991).

3 This article invokes the term *frontier* with the same spirit that has informed the work of western historians and environmental historians over the past ten years. Kerwin Lee Klein, "Reclaiming the 'F' Word, or Being and Becoming a Postwestern," *Pacific Historical Review* 65 (May 1996): 179–215; Stephen Aron, "Lessons in Conquest: Towards a Greater Western History," *Pacific Historical Review* 63 (May 1994): 125–48; David Wrobel, "Beyond the Frontier-Region Dichotomy," *Pacific Historical Review* 65 (August 1996): 401–29; Michael Steiner, "From Frontier to Region: FJT and the New Western History," *Pacific Historical Review* 64 (November 1996): 479–503; Richard White, Patricia Nelson Limerick, and James Grossman, *The Frontier in American Culture: An Exhibition at the Newberry Library, August 26, 1944–January 7, 1955* (Berkeley: University of California Press, 1994); and William Cronon, "The Trouble with Wilderness; or, Getting Back to the Wrong Nature," in *Uncommon Ground: Toward Reinventing Nature*, ed. William Cronon (New York: W. W. Norton, 1995), 69–90.

4 G. Edward White, *The Eastern Establishment and the Western Experience: The West of Frederick Remington, Theodore Roosevelt, and Owen Wister* (New Haven, Conn.: Yale University Press, 1968).

5 Richard Slotkin, *Gunfighter Nation: The Myth of the Frontier in Twentieth-Century America* (New York: Harper Perennial, 1993).

6 David Wrobel, *The End of American Exceptionalism: Frontier Anxiety from the Old West to the New Deal* (Lawrence: University Press of Kansas, 1993); also see Peter Schmitt, *Back to Nature: The Arcadian Myth in Urban America* (New York: Oxford University Press, 1969).

7 The literature on eastern hunters and environmental reform runs deep but has taken some dramatic turns of late: Andrew Isenberg, *The Destruction of the Bison: An Environmental History, 1750–1920* (New York: Cambridge University Press, 2000); Mark Spence, *Dispossessing the Wilderness: Indian Removal and the Making of the National Parks* (New York: Oxford University Press, 1999); Karl Jacoby, *Crimes against Nature: Squatters, Poachers, Thieves, and the Hidden History of American Conservation* (Berkeley: University of California Press, 2001); Louis Warren, *The Hunter's Game: Poachers and Conservationists in Twentieth-Century America* (New Haven, Conn.: Yale University Press, 1997).

8 Samuel Hays, *Conservation and the Gospel of Efficiency: The Progressive Conservation Movement, 1890–1920* (Cambridge: Harvard University Press, 1959).

9 Paul Lawrence Farber, *Finding Order in Nature: The Naturalist Tradition from Linnaeus to E. O. Wilson* (Baltimore: Johns Hopkins University Press, 2000), 88. Also see John Michael Kennedy, "Philanthropy and Science in New York City: The American Museum of Natural History" (Ph.D. diss., Yale University, 1968), 76–155; Douglas J.

Preston, *Dinosaurs in the Attic: An Excursion into the American Museum of Natural History* (New York: St. Martin's Press, 1986).

10 Donna Haraway, *Primate Visions: Gender, Race, and Nature in the World of Modern Science* (New York: Routledge, 1989).

11 Thomas Richards, *The Imperial Archive: Knowledge and the Fantasy of Empire* (New York: Verso, 1993). Richards focuses on the British Empire, but his framework is helpful for thinking about the wider social and political functions of museums in American culture, especially at the turn of the century.

12 "The Whale in the American Museum of Natural History," *Scientific American Supplement* 64 (September 14, 1907): 161–62.

13 To be clear, the practice of harvesting commodities from the ocean was in no way new to the twentieth century, and America's true heyday of ocean harvesting would not happen until after World War II. What is new to the early twentieth century is the cultural meaning of the whale. The mysterious monster of *Moby-Dick* was being demystified, demythologized, and incorporated into the machinery of Ford's, Taylor's, and Roosevelt's worlds.

14 Andrews to Dr. H. C. Hovey (April 10, 1908), Andrews Papers/DM, Folder VII-I, "Whale Collecting."

15 Frederick True, cetologist for the Smithsonian Institution, was probably the first American natural historian to make use of Newfoundland shore stations in his study of North Atlantic baleen whales.

16 Andrews to Captain Balcom (April 10, 1908), Andrews Papers/DM, Folder VII-I, "Whale Collecting."

17 In 1881 John Muir observed the whaling operations, though not shore stations, of the North Pacific and drew an obvious, but revealing, analogy to the American West: "Newly discovered whaling grounds, like gold mines, are soon overcrowded and worked out, the whales being either killed or driven away" (*The Cruise of the Corwin* [New York: Houghton Mifflin, 2000], 191–92). It is also worth mentioning that some of these stations have entered into a postindustrial phase in the form of tourism.

18 William Cronon, *Nature's Metropolis: Chicago and the Great West* (New York: W. W. Norton, 1991).

19 Richard Ellis, *Men and Whales* (New York: The Lyons Press, 1999), 337–85.

20 Roy Chapman Andrews, *Whale Hunting with Gun and Camera* (New York: Appleton & Co., 1916), 22–37.

21 Andrews to Mr. J. Quinton (September 16, 1908), Andrews Papers/DM, Folder VII-I, "Whale Collecting."

22 Roy Chapman Andrews, *Whale Hunting with Gun and Camera* (New York: D. Appleton, 1916), 49.

23 Roy Chapman Andrews, "Whale-Hunting as it is Now Done," *World's Work* 17 (December 1908): 11031; Gregg Mitman, *Reel Nature: America's Adventure with Wildlife on Film* (Cambridge: Harvard University Press, 1999), 5–25; Haraway, *Primate Visions*, 42–46.

24 Incidentally, this was not exactly true for all naturalists, especially those ship captains and surgeons in the business of whaling during the nineteenth century.

25 Andrews to Mr. A. Garcin (December 5, 1908), Andrews Papers/DM, Folder VII-I, "Whale Collecting."

26 "Camera Hunt for Whales," *New York Times,* September 14, 1908, 2; Roy Chapman Andrews, "Shore-Whaling: A World Industry," *National Geographic Magazine* 22 (May 1911): 413.

27 Andrews to J. Edmund Clark (July 21, 1911), Andrews Papers/DM, Folder VII-I, "Whale Collecting."

28 Letter from a stockholder of the Pacific Whaling Company to Andrews (February 9, 1909), Andrews Papers/DM, Folder VII-I, "Whale Collecting."

29 Andrews to stockholder of the Pacific Whaling Company (February 26, 1909), Andrews Papers/DM, Folder VII-I, "Whale Collecting."

30 Michael Robinson, *The Coldest Crucible: Arctic Exploration and American Culture* (Chicago: University of Chicago Press, 2006); Beau Riffenburgh, *The Myth of the Explorer: The Press, Sensationalism, and Geographical Discovery* (New York: Oxford University Press, 1994).

31 Andrews, *Under a Lucky Star,* 46; letter from Roy Chapman Andrews to the Players (November 20, 1913), Andrews Papers/DM, Folder VII-I, "Whale Collecting."

32 "Hunting Whales in the Far North with a Camera," *New York Times,* September 27, 1908, 1.

33 Joe Dubbert, "Progressivism and the Masculinity Crisis," in *The American Man,* ed. Elizabeth Pleck and Joseph Pleck (Englewood Cliffs, N.J.: Prentice-Hall, 1980), 303–20; Peter Filene, *Him/Her/Self: Sex Roles in Modern America,* 2nd ed. (Baltimore: Johns Hopkins University Press, 1986); Kim Townsend, *Manhood at Harvard: William James and Others* (New York: W. W. Norton, 1996).

34 "Hunting Whales in the Far North with a Camera," *New York Times,* September 27, 1908, 1.

35 Quoted in Geoffrey Hellman, *Bankers, Bones and Beetles: The First Century of the American Museum of Natural History* (Garden City, N.Y.: Natural History Press, 1968), 174–75.

36 Andrews, *Under a Lucky Star,* 135.

37 Andrews, "Shore-Whaling," 426; Andrews, *Whale Hunting,* 297. Andrews likely was aware of the term *commercial extinction* from Joel Allen. See Joel Allen, "The North Atlantic Right Whale and Its Near Allies," *Bulletin of the American Museum of Natural History* 24 (1908): 278.

38 Andrews would later meet officials from the Japanese Government Bureau of Science to go over whale statistics for composing a report on the future economics of whaling. He also exchanged several letters with J. Edmond Clark, a British official who was interested in regulating the whaling industry off the coast of South Africa.

39 Roy Chapman Andrews, "A Bill for the Protection of Whales and the Regulation of the Shore Whale Fishery in Alaska," n.d., Andrews Papers/DM, Folder VII-I, "Whale Collecting." Also see Andrews to J. Edmund Clark (July 21, 1911), Andrews Papers/DM, Folder VII-I, "Whale Collecting." Apparently Andrews never submitted the bill. The draft is undated but was probably written in 1911.

40 "To Protect Sea Mammals," *New York Times,* November 25, 1909, 4.

41 Kurkpatrick Dorsey, *The Dawn of Conservation Diplomacy: U.S.-Canadian Wildlife Protection Treaties in the Progressive Era* (Seattle: University of Washington Press, 1998).

42 "Closed Season for Whales," *New York Times,* June 30, 1912, 7.

43 Helen Rozwadowski, *The Sea Knows No Boundaries: A Century of Marine Science under ICES* (Copenhagen: International Council for the Exploration of the Sea, 2002).
44 "Threatening Marine Food Supply," *Christian Science Monitor,* June 1, 1912, 36.
45 "Conservation of Whales," *New York Times,* November 1, 1910, 3.
46 Frederic A. Lucas, "The Whale-Hunting Industry," *Scientific American Supplement* 65, no. 1671 (January 11, 1908): 30.
47 Frederic A. Lucas, "The Passing of the Whale," *Scientific American Supplement* 66, no. 1717 (November 28, 1908): 337, 344–46.
48 "Conservation of Whales," 8:3.
49 Theodore Roosevelt et al., "Scientific Surveys of the Philippine Islands," *Science,* n.s., 21 (May 19, 1905): 761–70.
50 Andrews, *Under a Lucky Star,* 52.
51 Andrews to John B. Trevor (September 5, 1911), Andrews Papers/DM, Folder III-1, "Expeditions."
52 Andrews to Allen (March 9, 1910), Andrews Papers/DM, Folder II-2, "Allen J."
53 Allen to Andrews (May 20, 1910), Andrews Papers/DM, Folder II-2, "Allen J."
54 For the paucity of specimens held by museums, see Allen, "North Atlantic Right Whale," 279.
55 Andrews to Allen (July 25, 1910), Andrews Papers/DM, Folder II-1, "Allen, J."
56 Andrews to C. H. Wells (January 11, 1911), Andrews Papers/DM, Folder VII-1, "Whale Collecting."
57 F. G. Aflalo, "Fishing vs. Shooting as a Remedy for Brain Fag," *Outing Magazine* 52, no. 4 (July 1908): 429–32. For Walton on hunting versus fishing, see Izaak Walton, *The Complete Angler, or The Contemplative Man's Recreation* (Boston: Little, Brown, 1912), 17–52.
58 A. W. Dimock, "King Tarpon, the High Leaper of the Sea," *Outing Magazine* 53 (March 1909): 703.
59 Mrs. Oliver C. Ginnell, *Introduction of American Big Game Fishing,* ed. Eugene V. Connett (New York: Derrydale Press, 1935), xiv–xv.
60 Ibid., xiv.
61 Moise N. Kaplan, *Big Game Fishermen's Paradise: A Complete Treatise* (Tallahassee, Fla.: Rose Printing, 1936), 282.
62 Kevin S. Blake, "Zane Grey and Images of the American West," *Geographical Review* 85, no. 2 (April 1995): 202–16.
63 Gifford Pinchot is another telling example. Gifford Pinchot, *To the South Seas* (Philadelphia: John C. Winston, 1930).
64 William Gregory, "Tour of the New Hall of Fishes," *Natural History* 28 (1928): 9–10; E. W. Gudger, "Capture of an Ocean Sunfish," *Scientific Monthly* 26 (March 1928): 257–61.
65 Andrews to Dr. Ephraim Cutter (May 13, 1915), Andrews Papers/DM, Folder VIII-3, "Whale Use as Food."
66 Andrews, *Whale Hunting,* 90.
67 Nancy Shoemaker, "Whale Meat in American History," *Environmental History* 10, no. 2 (April 2005): 269–94.
68 "Can Order a Whale Steak," *Washington Post,* January 12, 1908, M3.
69 "Whale Meat Called Venison of the Sea," *Christian Science Monitor,* June 27, 1917, 12; "Whale Meat and Fish Oddities," *New York Times,* October 5, 1918, 4.

70 "Whale Meat Lunch to Boost New Food," *New York Times,* February 9, 1918, 1.
71 Andrews, *Whale Hunting,* 86–87.
72 "The Praises of Whale Meat," *New York Times,* February 10, 1918, 4.
73 Herbert B. Nichols, "Mammoth Appetites Explore a . . . Mammoth," *Christian Science Monitor,* January 17, 1951; reprinted in *Explorers Journal* 29 (winter–spring 1951): 42–55.
74 "Whale Meat and Fish Oddities."
75 Carol J. Adams, *The Sexual Politics of Meat: A Feminist-Vegetarian Critical Theory* (New York: Continuum, 1990), 120–42.
76 "May Use Whale Meat in the Army Ration," *New York Times,* March 4, 1918, 2.
77 C. H. Claudy, "The Whale as a Food Factor," *Scientific American* 118 (March 9, 1918): 208–9.
78 The point is subtly made here, but is powerfully articulated by Richard Grove, *Green Imperialism: Colonial Expansion, Tropical Island Edens, and the Origins of Environmentalism, 1600–1860* (Cambridge: Cambridge University Press, 1995).
79 Also see "Leviathan," *Living Age* 252 (February 2, 1907): 312–13, for a description of how modern ship technology increased the depth and breadth of whale natural history.
80 Roy Chapman Andrews, "What Shore-Whaling is Doing for Science," *Nature* 88 (December 28, 1911): 281. On nineteenth-century cetology, see Lyndall Baker Landauer, "From Scoresby to Scammon: Nineteenth Century Whalers in the Foundations of Cetology" (Ph.D. diss., 1982).
81 For an intriguing analysis of the Taylorization of science, see Bonnie Tocher Clause, "The Wistar Rat as a Right Choice: Establishing Mammalian Standards and the Ideal of a Standardized Mammal," *Journal of the History of Biology* 26 (1993): 329–50.
82 Andrews was working under the guidance of the curator of mammals, Joel Allen, who held a similar belief in the differences between Atlantic and Pacific representatives of the right whale. See Allen, "North Atlantic Right Whale," 279.
83 Andrews to Victor H. Street (July 14, 1909), Andrews Papers/DM, Folder VII-I, "Whale Collecting."
84 Roy Chapman Andrews, "Explorers and Their Work," *Saturday Evening Post* 204 (August 22, 1931): 6–7, 84–85; Roy Chapman Andrews, *This Business of Exploring* (New York: G. P. Putnam's Sons, 1935).
85 There is the possibility that Andrews was responding to biology's embrace of experimental methods around the turn of the century, an episode in the history of biology sometimes dubbed the "revolt from morphology." As biology moved into the laboratory, as in Woods Hole's embryological research program or Morgan's *Drosophila* lab, museum-based natural historians often found themselves fighting for legitimacy. Andrews's modernization of natural history would then be interpreted as a defensive posture in a social and epistemological battle between experimentalists and natural historians. As appealing as such an explanation may be, Andrews never seems to have engaged in such debates. For an interesting window into the problems that museum naturalists faced at the turn of the century, see Lynn Nyhart, "Natural History and the 'New' Biology," in *Cultures of Natural History,* ed. Nicholas Jardine, J. A. Secord, and E. C. Spary (Cambridge: Cambridge University Press, 1996), 426–46.
86 Henry Fairfield Osborn to Andrews (July 7, 1914), Andrews Papers/DM, Folder II-5C, "Osborn re: 3rd Asiatic."

87 Andrews to J. A. C. Smith (November 111, 1915), Andrews Papers/DM, Folder III-1, "Expeditions."

CHAPTER 2: ROBERT CUSHMAN MURPHY AND THE NATURAL HISTORY OF OCEAN ISLANDS

1 Robert Cushman Murphy, *Logbook for Grace* (New York: Time Incorporated, 1965), 115.
2 Marson Bates, *The Nature of Natural History* (London: Chapman & Hall, 1951), 7.
3 Richard H. Pough, "The Museum and Natural Resources" (October 4, 1950), Murphy Papers, Folder "Pough, Dick."
4 Robert Cushman Murphy, "Conservation V," delivered to the Conservation Committee of the Garden Club of America on November 15, 1939; published in the Garden Club of America *Bulletin,* May 1940, 41.
5 Robert Cushman Murphy, "Greek and Latin in the College Curriculum" (June 1911), commencement address at Brown University, Murphy Papers, Folder "Murphy: Greek and Latin."
6 Murphy, *Logbook for Grace,* xxii–xxiv.
7 "Review of the Whaling Fishing Industry," New York Chamber of Commerce (1915); see pasted articles in Journal #41, "The Way of the Sperm Whaler," Murphy Papers.
8 Letters to the editor from J. A. Morch, "The Passing of the Whale," *Scientific American Supplement* 67 (March 13, 1909): 171; F. A. Lucas, "The Passing of the Whale," *Scientific American Supplement* 67 (April 24, 1909): 27; and J. A. Morch, "The Passing of the Whale," *Scientific American Supplement* 67 (June 12, 1909): 379.
9 Lucas to Andrews (May 16, 1911), and Andrews to Lucas (May 18, 1911), Andrews Papers/DM, Folder II-4 "Lucas."
10 Murphy's expedition to South Georgia has recently been elegantly told by Eleanor Mathews, *Ambassador to the Penguins: A Naturalist's Year Aboard a Yankee Whaleship* (Boston: David R. Godine, 2003).
11 Murphy, *Logbook for Grace,* 177.
12 Robert Cushman Murphy, "South Georgia, an Outpost of the Antarctic," *National Geographic* 41 (April 1922): 415. Also see Robert Cushman Murphy, "The Status of Sealing in the Sub-Antarctic Atlantic," *Scientific Monthly* 7 (August 1918): 112–19.
13 Murphy, *Logbook for Grace,* 180.
14 Robert Cushman Murphy, "The Way of the Sperm Whaler, Part III," *Sea Power* 3 (August 1917): 50–54; Robert Cushman Murphy, "Sub-Antarctic Whaling," *Sea Power* 3 (September 1917): 44–47.
15 Murphy, *Logbook for Grace,* 190.
16 Murphy, "Sub-Antarctic Whaling," 44.
17 Ibid.
18 Robert Cushman Murphy, "Lo, the Poor Whale," *Science* 91 (April 19, 1940): 373–76.
19 Joseph William Collins, "Decadence of the New England Deep-Sea Fisheries," *Harpers Monthly* 94 (March 1897): 610.
20 James Lindgren, "Let Us Idealize Old Types of Manhood: The New Bedford Whaling Museum, 1903–1941," *New England Quarterly* 72, no. 2 (June 1999): 163–206.

21 Murphy to Charles Davenport (June 9, 1936), Murphy Papers, Folder "Davenport."
22 Minutes of meeting, "The Four Whaling Scenes" (January 12, 1927), AMNH Special Collections, Exhibits Folder, Hall of Ocean Life.
23 H. E. Anthony, "Glimpses into the Hall of Ocean Life," *Natural History* 33 (July–August 1933): 365.
24 Form letter from "The Whaling Museum Society, Incorporated" to a Mr. Fisher (October 27, 1941), likely penned by Charles Davenport, Murphy Papers, Folder "Davenport."
25 Lewis Morris Iddingss, "The Art of Travel: Ocean Crossings," *Scribner's Monthly* 21 (April 1897): 425–47; Charles William Kennedy, "Gambling on Ocean Steamers," *North American* 150 (June 1890): 780–82; W. P. Reeves, "All Red Route," *Cornhill* 97 (February 1908): 176–88; E. A. Stevens, "Safety of Travel on the Modern Ocean Liner," *Review of Reviews* 39 (March 1909): 330–35; John H. Gould, "Ocean Passenger Travel," *Scribner's Magazine* 9 (April 1891): 393–418.
26 Joseph Conrad, "Ocean Travel," in *Last Essays*, ed. Richard Curle (Garden City, N.Y.: Doubleday, Page, 1926), 35–38.
27 Robert Cushman Murphy, "Avian Orders of the Tubinares" (M.S. thesis, Columbia University, 1918), 471, Murphy Papers, Folder "Murphy: Avian Order of Tubinares."
28 The somewhat counterintuitive fact that ocean productivity varies inversely with temperature was discovered by the German marine scientist Victor Hensen in 1889. Peter Bowler, *The Environmental Sciences* (New York: W. W. Norton, 1992), 386.
29 Murphy to Francisco Ballen (January 31, 1920) for quote; also Murphy to Mr. Loiseau (January 25, 1920); and Murphy to R. H. Eggleston (December 27, 1919), all pasted in the back on Murphy's bound expedition notes, Murphy Papers, Journal #12, "Peruvian Littoral Expedition, 1919–1920."
30 Robert Cushman Murphy, Journal #12, "Peruvian Littoral Expedition, 1919–1920," 81.
31 Murphy was building on the work of R. E. Coker, who had recently worked in the region. The main difference between their work seems to be Murphy's analysis of phytoplankton content. R. E. Coker, "Ocean Temperatures off the Coast of Peru," *Geographical Review* 5 (February 1918): 127–35; R. E. Coker, "Peru's Wealth-Producing Birds," *National Geographic Magazine* 37 (June 1920): 537–66; and R. E. Coker, "The Fisheries and the Guano Industry of Peru," Proceedings of the Fourth International Fishery Congress, *Bulletin of the Bureau of Fisheries* 28 (1908): 333–65.
32 Robert Cushman Murphy, "The Oceanography of the Peruvian Littoral with Reference to the Abundance and Distribution of Marine Life," *Geographical Review* 13 (January 1923): 71–72.
33 Ibid., 64–71.
34 Quoted in Robert Cushman Murphy, "The Progress of Science," *Scientific Monthly* 56 (June 1943): 570.
35 Robert Cushman Murphy, *Bird Islands of Peru: The Record of a Sojourn on the West Coast* (New York: G. P. Putnam's Sons, 1925), 23–35.
36 Murphy was oddly silent regarding developments in ecological biogeography in the 1920s and 1930s, though it seems that his work can be characterized as nothing but that. He did express, however, some misgivings with W. C. Allee and Karl P. Schmidt's 1937 rewritten edition of Richard Hesse, *Tiergeographie auf oekologischer Grundlage* (1924), mostly for their handling of oceanography. Robert Cushman Murphy, "Animal Geography: A Review," *Geographical Review* 28 (January 1938): 140–44.

37 Roy Chapman Andrews, then director of the AMNH, thought that Murphy was being a little too optimistic in requesting a run of some two thousand volumes. He only saw to printing five hundred. *Sea Birds of South America* sold out faster than any other publication in the museum's history. It was because of decisions like this that Andrews's directorship often raised the ire of museum naturalists.

38 Robert Cushman Murphy, *Oceanic Birds of South America* (New York: Macmillan, 1936), 1:59–60.

39 Murphy to Francisco Ballen (January 25, 1920), Murphy Papers, pasted in back of Journal #12, "Peruvian Littoral Expedition, 1919–1920."

40 Robert Cushman Murphy, "The Most Valuable Bird in the World," *National Geographic Magazine* 46 (September 1924): 278–302.

41 Gary Kroll, "The Pacific Science Board in Micronesia: Preparation and Preservation of a New Frontier Territory," *Minerva* 41, no. 1 (2003): 25–46.

42 Robert Cushman Murphy, Murphy Papers, Journal #28, "Pearl Island Expedition of 1945," 1–2, 24–25, 47.

43 Robert Cushman Murphy to Señor Francisco Ballén (July 21, 1938), Murphy Papers, "Ballén, Señor F."

44 Robert Cushman Murphy, "Whitney Wing," *Natural History* 44 (September 1939): 101.

45 Julia Norton Babson to Murphy (n.d.), Murphy Papers, Folder "Babson, Julia Norton."

46 Murphy to C. V. Whitney (June 2, 1939), 2, Murphy Papers, Folder "Whitney, Cornelius Vanderbilt 1."

47 Cornelius Whitney to Murphy (December 15, 1943), Murphy Papers, Folder "Whitney, C. V."

48 Robert Cushman Murphy, "Digest of the History, Status and Program of the Exhibits in Whitney Memorial Hall" (December 21, 1943), Murphy Papers, Folder "Murphy: Digest of the History. . . . "

49 Robert Cushman Murphy, Murphy Papers, Journal #16, "New Zealand Expedition," 10–11.

50 *New Zealand Herald* (September 1 and October 2, 1948).

51 Murphy, Murphy Papers, Journal #16, "New Zealand Expedition," 42.

52 Murphy to Albert Parr (May 6, 1948), Murphy Papers, pasted in back of Journal #16, "New Zealand Expedition."

53 Murphy, Murphy Papers, Journal #16, "New Zealand Expedition," 200.

54 Robert Cushman Murphy, "A New Zealand Expedition of the American Museum of Natural History," *Science* 108 (October 29, 1948): 463–64.

55 Murphy was fully aware that most moa species had taken their leave from this world even before New Zealand was populated by Europeans. Several species, however, remained and were hunted into extinction by European settlers.

56 Murphy to Albert Parr (May 6, 1948), Murphy Papers, pasted in back of Journal #16.

57 Robert Cushman Murphy, "The Seventh Pacific Science Congress," *Scientific Monthly* 69 (August 1949): 84–92.

58 Robert Cushman Murphy, "The Impact of Man upon Nature in New Zealand" (address to the Pacific Science Congress), November 5, 1958, printed in *Proceedings of the American Philosophical Society* 95 (December 1951): 570–72.

59 Ibid., 572–74.

60 Ibid., 577–82.

61 Robert Cushman Murphy, "The Need of Insular Exploration as Illustrated by Birds," *Science* 88 (December 9, 1938): 533–39.

62 Robert Cushman Murphy, Murphy Papers, Journal #42, "The Birds of Long Island," title page.

63 Grace E. Barstow Murphy, *There's Always Adventure: The Story of a Naturalist's Wife* (New York: Harper & Brothers, 1943), 28.

64 Richard Harmond, "Robert Cushman Murphy and Environmental Issues on Long Island," *Long Island Historical Journal* 8 (fall 1995): 76–82; Murphy to Stuart Gracey (January 6, 1950), Murphy Papers, Folder "Brookhaven, NY Trustees of the Freeholders"; Robert Cushman Murphy, "Mosquito Control through Marsh Drainage" (n.d.), Murphy Papers, Folder "Mosquito Control"; Murphy to Charles Jackson (September 22, 1939), Murphy Papers, Folder "Jackson, Halseted."

65 Robert Cushman Murphy, *Fish-Shape Paumanok* (Philadelphia: American Philosophical Society, 1964), 37.

66 Ibid., 1.

CHAPTER 3: SENSATIONAL MANAGEMENT

1 "First Broadcast" (September 1932), a preliminary draft of what was to be said during the broadcast, edited by William Beebe. William Beebe Papers, Department of Rare Books and Special Collections, Princeton University Library, Collection C0661 (hereafter Beebe Papers/PUL), Box 12, Folder 4.

2 "Bathysphere Dive 20" (September 22, 1932), transcript of communication between Beebe and staff, Beebe Papers/PUL, Box 12, Folder 4.

3 "First Broadcast," 10.

4 Lawrence Buell, *The Environmental Imagination: Thoreau, Nature Writing, and the Formation of American Culture* (Cambridge, Mass.: Harvard University Press, 1995); Don Scheese, *Nature Writing: The Pastoral Impulse in America* (New York: Simon & Schuster Macmillan, 1996); Peter Fritzell, *Nature Writing and America: Essays upon a Cultural Type* (Ames: Iowa State University Press, 1990); and Leo Marx, *The Machine in the Garden: Technology and the Pastoral Ideal in America* (London: Oxford University Press, 1965). On travel writing, see Casey Blanton, *Travel Writing: The Self and the World* (New York: Simon & Schuster, 1997); and Mary Louise Pratt, *Imperial Eyes: Travel Writing and Transculturation* (London: Routledge, 1992).

5 Michael Robinson, *The Coldest Crucible: Arctic Exploration and American Culture* (Chicago: University of Chicago Press, 2006).

6 For similar use of the sublime see Michael Smith, *Pacific Visions: California Scientists and the Environment, 1850–1915* (New Haven, Conn.: Yale University Press, 1987); David Nye, *American Technological Sublime* (Cambridge, Mass.: MIT Press, 1994).

7 William Beebe, "A Wilderness Laboratory," *Atlantic Monthly* 119 (May 1917): 628–29; reprinted in *Jungle Peace* (New York: H. Holt, 1918).

8 A retrospective history of the Department of Tropical Research, probably written in the late 1920s, gives a helpful glimpse into the patrons of the DTR. Beebe Papers/NYZS, DTR, General Records, Box 1, Folder "History of the DTR."

9 Ellery Sedgwick to Beebe (March 27, 1918), Beebe Papers/PUL, Box 15, Folder 2.

10 Theodore Roosevelt, review of *Jungle Peace* by William Beebe, *New York Times Book Review,* October 13, 1918, BR1; Ben Ray Redman, "Different Observers," review of *Edge of the Jungle* by William Beebe, *Nation* 114 (January 11, 1922): 47; Nicholas Roosevelt, "More Wonders of the Jungle in Volume by Beebe," review of *Jungle Days* by William Beebe, *New York Times Book Review,* July 5, 1925, BR3.

11 "William Beebe, Naturalist, Dies; Bathysphere Explorer was 84," *New York Times,* June 6, 1962.

12 William Beebe, "Weird Fish Drawn from Sargasso Sea," *New York Times,* March 8, 1925, 1:7.

13 Henry Fairfield Osborn to Harrison Williams (March 30, 1925), Beebe Papers/PUL, Box 16, Folder 7.

14 "*Arcturus* Sails for Sargasso Sea," *New York Times,* February 11, 1925, 14:2.

15 "Beebe Ship Silent 11 Days," *New York Times,* April 10, 1925, 1:3.

16 George Palmer Putnam to Beebe (April 14, 1925), Beebe Papers/PUL, Box 16, Folder 7.

17 George Palmer Putnam to Beebe (April 17, 1925), Beebe Papers/PUL, Box 16, Folder 7.

18 Naomi Oreskes, "Objectivity or Heroism? On the Invisibility of Women in Science," in *Science in the Field,* ed. Henrika Kuklick and Robert Kohler, *Osiris,* 2nd ser., 11 (1996): 87–113.

19 "Beebe and *Arcturus* Home with Marvels," *New York Times,* July 31, 1925, 1:4, 8. At least one writer used the occasion to comment on the resilience of myths in the face of scientific examination. M. B. Levick, "Old Myths Defy the Light of Science," *New York Times,* August 30, 1925, 4.11.

20 William Beebe, "Ocean Tells New Tales to Beebe," *New York Times,* August 9, 1925, 4:1.

21 Helen Rozwadowski, *Fathoming the Ocean: The Discovery and Exploration of the Deep Sea* (Cambridge, Mass.: Harvard University Press, 2005).

22 William Beebe, *The Arcturus Adventure: An Account of the New York Zoological Society's First Oceanographic Expedition* (New York: Harper and Row, 1981), 340.

23 William Beebe, "In Pursuit of an Elusive Sea," *New York Times Magazine,* April 19, 1925, 4:1.

24 Beebe, *Arcturus Adventure,* 348.

25 Ibid., 341–42.

26 Ibid., 133–35.

27 R. L. Duffus, "'Arcturus,' Whither Away?," *New York Times Book Review,* May 23, 1926, 1.

28 Henry Chester Tracey, *American Naturists* (New York: E. P. Dutton, 1930), 222–23. Tracey perceptively put his finger on the entire dynamic that I am arguing here. In a summary of Beebe's oceanic writings as of 1930, he notes that "without the slightest effort at sensation-mongering some quiet descriptions shape themselves into forms of arresting strangeness. . . . And that is how he avoids, as we said before, certain pitfalls. Since the truth is strange enough, he conveys it without condiments" (224–25).

29 Lewis S. Gannett, "Whither Mr. Beebe?," *Nation* 123 (September 8, 1926): 225.

30 Alexander Petrunkevitch, "An Explorer of Nature," *Yale Review* 16 (January 1927): 404–6.

31 Beebe to Madison Grant (April 1, 1929), Beebe Papers/PUL, Box 7, Folder 5.

32 See ibid.; and Beebe to Madison Grant (June 1, 1929), Beebe Papers/PUL, Box 7, Folder 5, where Beebe reported that two parliamentary visitors "expressed themselves as being dissatisfied only if we do not go on for another year."

33 Beebe to Mr. Niles (August 1931), Beebe Papers/PUL, Box 7, Folder 5.

34 Beebe to Mr. Niles (May 1931), Beebe Papers/PUL, Box 7, Folder 5.

35 Beebe to Mr. Niles (July 1931), Beebe Papers/PUL, Box 7, Folder 5.

36 "The Beebe Expedition," *Royal Gazette Bermuda,* June 11, 1929; see Beebe Papers/NYZS, DTR, General Records, Box 3, Folder "G. Hollister Files."

37 Brad Matsen, *Descent: The Heroic Discovery of the Abyss* (New York: Pantheon Books, 2005).

38 "Deep Down," in "Notes on a Week's Headlines," *New York Times,* June 22, 1930, III, 7:5.

39 E. J. Allen to Beebe (January 21, 1931), Beebe Papers/PUL, Box 15, Folder 1.

40 Draft of encyclopedia article written by John T. Nichols. See Nichols to Beebe (March 2, 1931), Beebe Papers/PUL, Box 15, Folder 1.

41 C. Carl Borth, "A Modern Marco Polo," *Bermudian,* November 1931, 12.

42 Joseph J. Corn, *The Winged Gospel: America's Romance with Aviation, 1900–1950* (New York: Oxford University Press, 1983), 3–27.

43 Waldemar Kaempffert, "Into the Black Deeps of the Sea," *New York Times Magazine,* August 28, 1932, 8.

44 Cable from Philip Carlin of NBC to Beebe (September 22, 1932); cable from William Burke Miller, director of special broadcasts, NBC, to Beebe (September 22, 1932), Beebe Papers/PUL, Box 12, Folder 5. "Half a Mile Under the Sea," (London) *News Chronicle,* September 24, 1932. "The Loudspeaker," *New York American,* January 2, 1933. William Duncan, "Life a Mile Down in the Sea Described by Dr. William Beebe," (Philadelphia, Pa.) *Public Ledger,* January 18, 1933.

45 William Beebe, "Descent into Perpetual Night," *New York Times Magazine,* October 9, 1932, 1–2, 14–15.

46 Gilbert Grosvenor to Beebe (December 20, 1933), Beebe Papers/PUL, Box 17, Folder 4.

47 Beebe to Madison Grant (December 28, 1933), Beebe Papers/PUL, Box 15, Folder 15.

48 Reid Blair to Madison Grant (ca. January 1934), quoted in William Bridges, *Gathering of Animals: An Unconventional History of the New York Zoological Society* (New York: Harper & Row, 1974), 428.

49 Beebe to Dr. LaGorce (August 17, 1934), Beebe Papers/PUL, Box 17, Folder 4.

50 Telegram from Fisher to Beebe (July 9, 1934), Beebe Papers/PUL, Box 17, Folder 4.

51 "Beebe Wires Amazing Finds to Geographic," *Washington Herald* (August 12, 1934).

52 William Beebe, John Tee-Van, Gloria Hollister, Jocelyn Crane, and Otis Barton, *Half-Mile Down* (New York: Harcourt, Brace, 1934), 213.

53 "World Record Deep Sea Dives Made off Bermuda," *Geographic News Bulletin,* August 17, 1934, Beebe Papers/PUL, Box 17, Folder 4.

54 "Beebe Starts Descent to Half-Mile Sea Depth," (Philadelphia, Pa.) *Evening Bulletin,* August 11, 1934, 1.

55 Blair to Beebe (ca. 1934), quoted in Bridges, *Gathering of Animals*, 428–29.

56 William Beebe, "Bathysphere Observations Wanted" (ca. 1932), Beebe Papers/PUL, Box 12, Folder 4.

57 E. O. Hulbert to Beebe (March 31, 1934), Beebe Papers/PUL, Box 17, Folder 3.

58 "Beebe Out to Set Sea Descent Mark," *New York Times*, April 15, 1934.

59 Hugh Darby, review of *A Half-Mile Down* by William Beebe et al., *Nation* 139 (December 12, 1934): 687.

60 Carl Hubbs, "Reviews and Comments," *Copeia* 2 (July 16, 1935): 105.

61 John T. Nichols, "Life in the Bathysphere," *Saturday Review of Literature* 11 (December 8, 1934): 336.

62 "Dr. Beebe Returns from his 20th Trip," *New York Times*, November 3, 1934, 17:2.

63 On mountains and the geological sublime, see Marjorie Hope Nicolson, *Mountain Gloom and Mountain Glory: The Development of the Aesthetics of the Infinite* (Ithaca: Cornell University Press, 1959). For America's variants on the sublime, see Raymond O'Brien, *American Sublime: Landscape and Scenery of the Lower Hudson Valley* (New York: Columbia University Pres, 1981); Elizabeth McKinsey, *Niagara Falls: Icon of the American Sublime* (Cambridge: Cambridge University Press, 1985); Barbara Novak, *Nature and Culture: American Landscape and Painting, 1825–1875* (Oxford: Oxford University Press, 1980); Rob Wilson, *American Sublime: The Genealogy of a Poetic Genre* (Madison: University of Wisconsin Press, 1991). On the technological sublime, see Nye, *American Technological Sublime.* Earlier, and now classic, studies of the sublime came from scholars in America studies, including Marx, *Machine in the Garden;* and Roderick Nash, *Wilderness and the American Mind* (New Haven, Conn.: Yale University Press, 1967). For a critique of the sublime and environmentalism, see William Cronon, "The Trouble with Wilderness; or, Getting Back to the Wrong Nature," in *Uncommon Ground: Toward Reinventing Nature*, ed. William Cronon (New York: W. W. Norton, 1995), 69–90.

64 Sigmund Freud, *Civilization and its Discontents* (New York: W.W. Norton & Co., 1961), 11.

65 J. W. Van Dervoort, *The Water World, or The Ocean, Its Laws, Currents, Tides, Wind-Waves, Phenomena, Mechanical Appliances, Animal and Vegetable Life* (New York: Union Publishing House, 1883), 5.

66 William Beebe, "A Quarter Mile Down in the Open Sea," *Bulletin: New York Zoological Society* 33 (November–December 1930): 223.

67 Beebe et al., *Half-Mile Down*, 154.

68 William Beebe, "The Bermuda Oceanographic Expedition," *Bulletin: New York Zoological Society* 33 (March–April 1930): 61.

69 Beebe, "Quarter Mile Down," 208; William Beebe, "Down into Davy Jones's Locker," *New York Times Magazine*, July 13, 1930, 1.

70 Beebe, "Quarter Mile Down," 215.

71 William Beebe, "Dr. Beebe Again Will Invade Davy Jones's Locker," press release from National Geographic Society (ca. June 1934), 7, Beebe Papers/NYZS, DTR General, 1900–1962, scrapbook of Gloria Hollister. For a cultural history of space, see Howard E. McCurdy, *Space and the American Imagination* (Washington: Smithsonian Institution Press, 1997), 1–28.

72 Rudyard Kipling to Beebe (December 26, 1934), Beebe Papers/PUL, Box 15, Folder 17.

73 Beebe, "Quarter Mile Down," 201.
74 Ibid., 217.
75 Beebe et al., *Half-Mile Down,* 135.
76 William Beebe, "Resume" (ca. 1936), 1, 4, Beebe Papers/PUL, Box 12, Folder 8.
77 Spencer Tinker and Marian Omura, *Directory of the Public Aquaria of the World* (Honolulu: University of Hawaii, 1963), 4.
78 Charles Haskins Townsend, *The Public Aquarium: Its Construction, Equipment, and Management,* Bureau of Fisheries Document 1045 (Washington, D.C.: Government Printing Office, 1928), 250.
79 Gregg Mitman, *Reel Nature: America's Romance with Wildlife on Film* (Cambridge, Mass.: Harvard University Press, 1999), 159–65.
80 Barton was not alone in hoping to bring the underwater realm to the silver screen. In 1934, MGM began production of Jules Verne's *20,000 Leagues under the Sea.* It is not clear whether the idea of the picture was caused by the bathysphere publicity, but at least one newspaper writer did make the connection. Mayme Ober Peak, "Reel Life in Hollywood," (Boston, Mass.) *Globe,* November 26, 1934.
81 Beebe to Henry Blair (September 26, 1938), Beebe Papers/PUL, Box 15, Folder B.
82 William Beebe, review of *Titans of the Deep,* film by Otis Barton, in *Science,* 89, April 7, 1939, 319.
83 Robert W. Rydell, *World of Fairs: The Century-of-Progress Expositions* (Chicago: University of Chicago Press, 1993), 92–114.
84 Ibid., 111.
85 John Tee-Van, "The Zoological Society's Building at the World's Fair in 1940" (ca. 1940), Beebe Papers/NYZS, DTR, Office of the Director and General Associate.
86 Notes to speech at the 1944 New York Zoological Society Annual Dinner (ca. 1944), Beebe Papers/PUL, Box 14, Folder "Writings: Miscellaneous and Bibliographical."
87 William Soskin, "Reading and Writing," review of *Nonsuch: Land of Water* by William Beebe, *New York Evening News* (August 23, 1932), 25.
88 William Beebe, *Beneath Tropical Seas* (New York: Blue Ribbon Books, 1928), 3–6. Also see Beebe et al., *Half-Mile Down,* 3–19.

CHAPTER 4: RACHEL CARSON'S *THE SEA AROUND US*

1 "Whale Sighted in Sound at 5 A.M., and Blow Me Down Mates," *New York Times,* October 22, 1946, 26.
2 Ruth Adler, "'Whale Off!'—Manhattan," *New York Times,* November 3, 1946, 146.
3 The only earlier instance I have located was a mass beaching of eleven blackfish (pilot whales) at Kitty Hawk Beach Station in 1943; five off-duty coast guardsmen were able to haul to safety—by muscle and rope—at least one of the whales. "Coast Guard to the Rescue," *Christian Science Monitor,* September 23, 1943, 19.
4 E. R. Coker, *This Great and Wide Sea* (Chapel Hill: University of North Carolina Press, 1947), v.
5 Linda Lear, *Rachel Carson: Witness for Nature* (New York: Henry Holt, 1997), 61.
6 Rachel Carson to William Beebe (September 6, 1948), Carson Papers, Box 4, Folder 67.

7 Keith R. Benson, "From Museum Research to Laboratory Research: The Transformation of Natural History into Academic Biology," 49–86, and Philip Pauly, "Summer Resort and Scientific Discipline: Woods Hole and the Structure of American Biology, 1882–1925," 121–50, in *The American Development of Biology,* ed. Ronald Rainger, Keth Benson, and Jane Maienschein (New Brunswick, N.J.: Rutgers University Press, 1991).

8 Garland Allen, "Old Wine in New Bottles: From Eugenics to Population Control in the Work of Raymond Pearl," in *The Expansion of American Biology,* ed. Keith Benson, Jane Maienschein, and Ronald Rainger (New Brunswick, N.J.: Rutgers University Press, 1991), 231–61.

9 Paul Brooks, *The House of Life: Rachel Carson at Work, with Selections from Her Writings* (Boston: Houghton Mifflin, 1972), 125.

10 Linda Lear, too, suggests that this was a critical text in the evolution of Carson's writing; *Rachel Carson,* 82.

11 Rachel Carson, "Undersea," quoted in Linda Lear, ed., *Lost Woods: The Discovered Writing of Rachel Carson* (Boston: Beacon Press, 1998), 4 (originally published in *Atlantic Monthly* 160 [September 1937]: 322–25). Cf. William Beebe, *Beneath Tropical Seas* (New York: Blue Ribbon Books, 1932), 3–6; and William Beebe, John Tee-Van, Gloria Hollister, Jocelyn Crane, and Otis Barton, *Half-Mile Down* (New York: Harcourt, Brace, 1934), 3–19.

12 Carson, "Undersea," 6. Cf. William Beebe, *The Arcturus Adventure: An Account of the New York Zoological Society's First Oceanographic Expedition* (New York: Harper and Row, 1981), 194–219.

13 Carson, "Undersea," 8.

14 Ibid.

15 Ibid., 11.

16 Joan Shelly Rubin, *The Making of Middlebrow Culture* (Chapel Hill: University of North Carolina Press, 1992), 217. Rachel Carson to Henrik van Loon (June 3, 1939), Carson Papers, Box 3, Folder 57.

17 Rachel Carson to Henrik van Loon (February 5, 1938), Carson Papers, Box 3, Folder 57.

18 Samuel J. Looker, introduction to collection of essays edited by same, *Jefferies' England: Nature Essays by Richard Jeffries* (New York: Harper and Brothers Publishers, 1938), xi–xvii.

19 Rachel Carson to Dorothy Freeman (March 14, 1961), quoted in Martha Freeman, ed., *Always, Rachel: The Letters of Rachel Carson and Dorothy Freeman, 1952–1964* (Boston: Beacon Press, 1995), 360. Lear, *Rachel Carson,* 103–4.

20 Fred D. Crawford, *H. M. Tomlinson* (Boston: Twayne Publishers, 1981), 30.

21 Carson to Freeman (November 7, 1957), quoted in Freeman, *Always, Rachel,* 233.

22 Henry Betson, "The Outermost House," in *Especially Maine: The Natural World of Henry Betson from Cape Cod to the St. Lawrence,* ed. Elizabeth Coatsworth (Brattleboro, Vt.: Stephen Greene Press, 1970), 18.

23 On Williamson's influence on Carson, see Lear, *Rachel Carson,* 90–91.

24 Carson to Weeks (June 24, 1939), Lear-Carson Collection, Box 4, Folder "Origins of *Under the Sea-Wind.*"

25 Rachel Carson, *Under the Sea-Wind: A Naturalist's Picture of Ocean Life,* 2nd ed. (New York: Oxford University Press, 1952), 200.

26 "A Dramatic Picture of Ocean Life," *New York Times,* November 23, 1941, 10; George Miksch Sutton, "Along the Moonlit Tidal Flats," *Books* (December 14, 1941), 5; William Beebe, review of *Under the Sea-Wind, Saturday Review of Literature,* December 27, 1941, 5.

27 See Carson to Freeman (February 13, 1954), quoted in Freeman, *Always, Rachel,* 23.

28 Lear, *Rachel Carson,* 104.

29 It is a small tragedy that we do not have a good history of the typical bureau biologist. Until we do, see Kurkpatrick Dorsey's treatment of David Starr Jordan in *The Dawn of Conservation Diplomacy: U.S.-Canadian Wildlife Protection Treaties in the Progressive Era* (Seattle: University of Washington Press, 1998).

30 Lear, *Rachel Carson,* 82–83.

31 Rachel Carson, *Guarding Our Wildlife Resources, Conservation in Action,* no. 5. (Washington, D.C.: U.S. Fish and Wildlife Service, Government Printing Office, 1947), 2.

32 Bruce V. Lewenstein, "Public Understanding of Science in America, 1945–1965" (Ph.D. diss., University of Pennsylvania, 1987); Dorothy Nelkin, *Selling Science: How the Press Covers Science and Technology* (New York: W. H. Freeman, 1987), 86–91; Ronald C. Tobey, *The American Ideology of National Science, 1919–1930* (Pittsburgh: University of Pittsburgh Press, 1971), 62–95; Marcel C. Lafollette, *Making Science Our Own: Public Images of Science, 1910–1955* (Chicago: University of Chicago Press, 1990), 45–65.

33 Rachel Carson, "Science Keeps Watch over the Sea," draft of "Numbering Fish," for the *Baltimore Sun* (ca. 1936), Carson Papers, Box 98, Folder 1784.

34 Rachel Carson, "Sentiment Plays No Part in the Save-the-Shad Movement . . . ," in *Providence Sunday Journal,* February 28, 1937, Carson Papers, Box 98, Folder 1791.

35 Ralph Lutts, *The Nature Fakers: Wildlife, Science and Sentiment* (Golden, Colo.: Fulcrum Publishing, 1990), 172–73.

36 Rachel Carson, "Chesapeake Eels Seek the Sargasso Sea," *Baltimore Sunday Sun* (October 9, 1938); republished in Lear, *Lost Woods,* 19–23.

37 Rachel Carson, "Ace of Nature's Aviators," in Lear, *Lost Woods,* 24–29; Rachel Carson, "Sky Dwellers," *Coronet,* November 1945; Rachel Carson, "The Bat Knew It First," *Collier's,* November 18, 1944.

38 Rachel Carson to William Beebe (October 26, 1945), Carson Papers, Box 4, Folder 67.

39 Rachel Carson to DeWitt, editor of *Readers Digest,* May 31, 1945, Carson Papers, Box 98, Folder 1787.

40 Rachel Carson to William Beebe (September 6, 1948), Carson Papers, Box 4, Folder 67.

41 Edwin Way Teale, *North with the Spring: A Naturalist's Record of a 17,000-Mile Journey with the North-American Spring* (New York: Dodd, Meade, 1951), 2.

42 Lear, *Rachel Carson,* 141. One telling clipping in Carson's notes for the project is a "Topics of the Times," *New York Times,* June 5, 1949, E10.

43 Rachel Carson, *The Sea Around Us* (New York: Oxford University Press, 1951), vi.

44 William Beebe, ed., *The Book of Naturalists: An Anthology of the Best Natural History* (New York: Alfred A. Knopf, 1944).

45 Carson to Beebe (August 26, 1949) and Beebe to Carson (December 11, 1951), Carson Papers, Box 4, Folder 67.

46 Carson to Beebe (April 5, 1949), Carson Papers, Box 4, Folder 67.

47 Carson's information here comes largely from J. W. Gregory, "Geological History of the Pacific Ocean," *Proceedings of the Geological Society of London* 86 (1930).

48 Carson, *The Sea Around Us,* 7.
49 Ibid., 15.
50 Unsigned review, (Omaha, Neb.) *World Herald,* September, 2, 1951. See also Carson papers, Box 9, Folder 163.
51 "Great Seas Are Our Life," *Daily Oklahoman,* September 16, 1951, Carson Papers, Box 9, Folder 163.
52 "Man's Attempt to Subdue the Sea," (New Haven, Conn.) *Independent,* August 11, 1956.
53 Bruce Barton, "Bruce Barton Says," *Miami Herald,* September 9, 1951, Carson Papers, Box 9, Folder 163.
54 Donald Adams, "Speaking of Books," *New York Times Book Review,* March 22, 1953: 2.
55 Carol B. Gartner, "When Science Writing Becomes Literary Art," in *And No Birds Sing: Rhetorical Analyses of Rachel Carson's "Silent Spring,"* ed. Craig Waddell (Carbondale: Southern Illinois University Press, 2000), 105.
56 Carson even noted to Henry Bigelow, while researching the book, that "mathematicians do not help me much." Carson to Bigelow (March 4, 1950), Carson Papers, Box 4, Folder 68.
57 Carson to Freeman (November 5, 1957), quoted in Freeman, *Always, Rachel,* 232.
58 Carson, *The Sea Around Us,* 83.
59 Beebe, *Arcturus Adventure,* 317–38; Carson, *The Sea Around Us,* 89–92.
60 Carson, *The Sea Around Us,* 92.
61 See Carson's notes on Pacific research in her spiral notebook, Carson Papers, Box 4, Folder 78.
62 Carson, *The Sea Around Us,* 95–96. Carson's primary contact on the Pacific War Memorial was Robert Cushman Murphy. The line on "natural museums" first appeared in a letter to Murphy as a question that Carson wanted confirmed. This is typical of her research. Many notebooks are filled with questions that arose in Carson's mind, and she would then interview scientists or send letters asking for information. Carson to Murphy (October 23, 1848, and September 22, 1948), Carson Papers, Box 4, Folder 73.
63 Carson, *The Sea Around Us,* 215–16.
64 J. S. Colman, review of *The Sea Around Us* by Rachel Carson, *New Republic* 125 (August 20, 1951): 20.
65 Graham Netting, "The Naturalist's Bookshelf," *Carnegie Magazine* 25 (November 1951): 320.
66 Harvey Breit, review of *The Sea Around Us* by Rachel Carson, *Atlantic* 188 (August 1951): 84. A condensed and serialized portion of *The Sea Around Us* had appeared in the *New Yorker* shortly before the book's publication. It was the first nonhuman subject ever to be published in the Profiles section of the magazine.
67 Callum Roberts, *The Unnatural History of the Sea* (Washington: Island Press, 2007), 167–68.
68 Beebe, *Arcturus Adventure,* 201.
69 John Bardach, *Harvest of the Sea* (New York: Harper and Row, 1968), 5.
70 William Vogt, *Road to Survival* (New York: William Sloane Associates, 1948), 285; Fairfield Osborn, *Our Plundered Planet* (New York: Grosset & Dunlap, 1948). On Osborn, see Andrew Jamison and Ron Eyerman, *Seeds of the Sixties* (Berkeley: University of California Press, 1994), 64–82; and Gregg Mitman, "When Nature Is the Zoo," in

Science in the Field, ed. Henrika Kuklick and Robert Kohler, *Osiris,* 2nd ser., 11 (1996): 121–33.

71 See notecard, "The Allan Hancock Foundation," in Carson Papers, Box 4, Folder 79; William Laurence, "Sea Soon May Yield Great Food Stores," *New York Times,* June 21, 1948, 1; "Alga May Avert Famine," *Science News Letter,* January 1, 1949; Waldemar Kaempffert, "Future Generations from the Sea," *New York Times,* October 23, 1949, sec. IV, 9:6.

72 Gordon A. Riley, "Food from the Sea," *Scientific American* 181 (October 1949): 16–19; Daniel Merriman, "Food Shortages and the Sea," *Yale Review* 39 (March 1950): 430–44; and "Topics of the Times," *New York Times,* July 17, 1952, 22.

73 Rachel Carson, "The Ocean and a Hungry World," draft of unpublished chapter. Carson Papers, Box 7, Folder 134.

74 Mary McCay, *Rachel Carson* (New York: Twayne Publishers, 1993), 50–51.

75 From Carson's application to the Eugene F. Saxton Foundation (May 1, 1949); quoted in Lear, *Rachel Carson,* 163.

76 Rachel Carson to Shirley Collier, lawyer at RKO, November 10, 1952, RCC, Box 11, Folder 193.

77 Rachel Carson, "Preface to the Second Edition of *The Sea Around Us,*" republished in Lear, *Lost Woods,* 106–7.

78 Ralph H. Lutts, "Chemical Fallout: *Silent Spring,* Radioactive Fallout, and the Environmental Movement," in *And No Birds Sing: Rhetorical Analyses of Rachel Carson's "Silent Spring,"* ed. Craig Waddell (Carbondale: Southern Illinois University Press, 2000), 33–37.

79 Rachel Carson, "Design for Nature Writing" speech at reception of Burroughs Award (April 1952), printed in Lear, *Lost Woods,* 94–95.

80 National Book Award acceptance speech for *The Sea Around Us,* delivered January 29, 1951, Rachel Carson Collection, Box 101, Folder 1883.

81 Carson to Beebe (November 3, 1950, and April 26, 1951), Carson Papers, Box 4, Folder 67.

82 Rachel Carson, *Edge of the Sea* (New York: Houghton Mifflin, 1955), 7.

CHAPTER 5: EUGENIE CLARK AND POSTWAR OCEAN ICHTHYOLOGY

1 Quoted in Eugene Balon, "The Life and Work of Eugenie Clark: Devoted to Diving and Science," *Environmental Biology of Fishes* 41 (1994): 89.

2 "Career Women," *Cornet* 33 (April 1953): 115–28.

3 Roger Angell, "World of Women, USA: Lady with a Spear," *Holiday* 17 (January 1955): 48–51.

4 Carolyn Anspacher, "Skin Dive, Mother-to-Be Told," *San Francisco Chronicle* (February 13, 1961), 36.

5 Margaret Rossiter, *Women Scientists in America: Struggles and Strategies to 1940* (Baltimore: Johns Hopkins University Press, 1982), 303; on gender and natural history, see 73.

6 Roy Chapman Andrews, "Explorers and Their Work," *Saturday Evening Post* 204 (August 22, 1931): 7.

7 See Marcia Myers Bonta, *Women in the Field: America's Pioneering Women Naturalists* (College Station: Texas A&M University Press, 1991); and Vera Norwood, *Made from This Earth: American Women and Nature* (Chapel Hill: University of North Carolina Press, 1993).
8 Joseph Turner, "Science for the Misses," *Science* 129 (March 20, 1959): 749.
9 Margaret Rossiter, *Women Scientists in America before Affirmative Action, 1940–1972* (Baltimore: Johns Hopkins University Press, 1995), 41–47. Also see Vivian Gornick, *Women in Science: Portraits from a World in Transition* (New York: Simon and Schuster, 1983), 120
10 Elaine Tyler May, "Rosie the Riveter Gets Married," in *The War in American Culture: Society and Consciousness during World War II*, ed. Lary May (Chicago: University of Chicago Press, 1996), 128–43.
11 Elaine Tyler May, *Homeward Bound: American Families in the Cold War Era* (New York: Harper Collins, 1988); Susan M. Hartmann, *The Home Front and Beyond: American Women in the 1940s* (Boston: Twayne Publishers, 1982); William Chafe, *The Paradox of Change: American Women in the 20th Century* (New York: Oxford University Press, 1991); Karen Anderson, *Wartime Women: Sex Roles, Family Relations and the Status of Women during World War II* (Westport, Conn.: Greenwood Press, 1981); William M. Turtle Jr., *"Daddy's Gone to War": The Second World War in the Lives of America's Children* (New York: Oxford University Press, 1993); Winifred D. Wandersee, *Woman's Work and Family Values, 1920–1940* (Cambridge, Mass.: Harvard University Press, 1981).
12 Gregg Mitman, *Reel Nature: America's Romance with Wildlife on Film* (Cambridge, Mass.: Harvard University Press, 1999); Alexander Wilson, *The Culture of Nature: North American Landscape from Disney to the Exxon Valdez* (Cambridge, Mass.: Blackwell Publishers, 1992), 53–116.
13 Eugenie Clark, *Lady with a Spear* (New York: Harper & Brothers, 1953), 11.
14 Balon, "Life and Work of Eugenie Clark," 122–23.
15 C. M. Breder Jr. and Eugenie Clark, "A Contribution to the Visceral Anatomy, Development, and Relationships of the Plectognathi," *Bulletin of the American Museum of Natural History* 88, article 5 (1947): 287–320.
16 Clark, *Lady with a Spear*, 41–42.
17 Ibid., 42. By the time Clark arrived there, AMNH's Department of Experimental Biology had been the site for experiments on the physiology and psychology of reproduction in the lower vertebrates. See Mitman, *Reel Nature*, 61–66.
18 Clark, *Lady with a Spear*, 46.
19 Eugenie Clark, "A Method for Artificial Insemination in Viviparous Fishes," *Science* 112 (December 15, 1950): 722–23.
20 Clark, *Lady with a Spear*, 48.
21 Eugenie Clark, acceptance speech for the Society of Woman Geographers Gold Medal, *Bulletin for the Society of Woman Geographers* (fall 1975): 77.
22 Gary Kroll, "The Pacific Science Board in Micronesia: Preparation and Preservation of a New Frontier Territory," *Minerva* 41 (2003): 25–46.
23 Clark, *Lady with a Spear*, 78.
24 Ibid., 98.
25 Ibid., 100–103.

26 Ibid., 109–10.
27 Ibid., 113–14.
28 Eugenie Clark, "Reef Fish Studies in the South Pacific," *Scientific Investigations in Micronesia #1* (Washington, D.C.: National Research Council, 1949); NRC-PSB archives, Folder "Sim Report #1." Chandra Mukerji, *A Fragile Power: Scientists and the State* (Princeton, N.J.: Princeton University Press, 1989).
29 Clark, *Lady with a Spear,* 192. See below for Clark's work on the functional hermaphroditism of certain flounders. The *Newsweek* reviewer of *Lady with a Spear* makes special note of the Kinsey comment and the child-bearing pipefish. "Fish Lady," *Newsweek* 42 (July 20, 1953): 102–3.
30 She was also the subject of several popular but short profiles that discussed her work in the Pacific and the Red Sea. "Red Sea Swimmer," *Time* 58 (November 19, 1951): 68; "Fish Lady," *New Yorker* 28 (March 8, 1952): 24–25.
31 Reprinted as "Wonders of the Deep, by Eugenie Clark," *Cairo Calling* 847, pt. 2 (June 9, 1951): 12–13.
32 Eugenie Clark, "Field Trip to the South Seas," *Natural History* 60 (1951): 8–16; idem, "The Lost Quarry," *Natural History* 61 (1952): 258–63; idem, "Siakong: Spear-fisherman Pre-eminent," *Natural History* 62 (1953): 224–34.
33 "Undersea Armor," *Newsweek* 27 (January 21, 1946): 60; John Tassen, "Tourists in the Underwater World," *New York Times Magazine,* June 27, 1954, 18.
34 Thomas Griffin, "Out of This World," *Collier's* (February 2, 1946), 24. Leslie Lieber, "The World's Most Dangerous Classroom," *Science Digest* 20 (December 1946): 84–87.
35 "Diving School," *Life* 25 (July 12, 1948): 45; "Underwater Campus," *American Magazine* 147 (February 1949): 97.
36 George Kent, "Man's Newest and Loveliest Adventure," *Reader's Digest* 64 (March 1954): 109–12.
37 John Tassen, "Tourists," 27.
38 David Bradley, *No Place to Hide* (Boston: Little, Brown, 1948), 128–30. On *No Place to Hide,* see Paul Boyer, *By the Bomb's Early Light: American Thought and Culture at the Dawn of the Atomic Age* (Chapel Hill: University of North Carolina Press, 1994), 82–92.
39 Eugenie Clark, review of *Diving to Adventure* by Hans Haas, *Natural History* 61 (January 1952): 4.
40 Richard Hubler, "Between the Devilfish and the Deep Blue Sea," *Collier's* 131 (January 24, 1953): 40.
41 Ibid., 41.
42 Kent, "Man's Newest and Loveliest Adventure," 110.
43 Tassen, "Tourists," 18.
44 Harry Shershow, "Fun Under Water," *Popular Science Monthly* 148 (April 1946): 114.
45 See Victor Boesen, "Adventures of a Shark Diver," *Saturday Evening Post* 224 (July 14, 1951): 18.
46 Pierre de Latil and Mary Thayer Muller, "Wonderland of Deep Sea Hunters," *Coronet* 32 (September 1952): 42.
47 James Jones, "Scuba Diving's Growing Popularity" (1960), unpublished manuscript, University of Minnesota Rare Book Library.
48 Clark, "Field Trip to the South Seas," 10.

49 Clifton Fadiman, "A Courageous Young Scientist Explores the Undersea Universe in *Lady with a Spear*," *Book-of-the-Month Club News*, June 1953, 2–3.
50 Gilbert Klingel, "Underwater Boswell," review of *Lady with a Spear, New York Times Book Review*, July 19, 1953, 3.
51 Lewis Gannett, "This Wonderful and Fishy World," review of *Lady with a Spear* by Eugenie Clark, in *New York Herald Tribune Book Review* (July 19, 1953), 1.
52 "Girl Under Water," review of *Lady with a Spear*, in *Nation* 177 (September 5, 1953): 197.
53 Balon, "Life and Work of Eugenie Clark," 124.
54 Eugenie Clark, *Lady and the Sharks* (Sarasota, Fla.: Mote Marine Laboratory, 1969), 35.
55 Eugenie Clark, "Mating of Groupers," *Natural History* 74 (June 1965): 22–25; Clark, "Functional Hermaphroditism and Self-Fertilization in a Serranid Fish," *Science* 129 (January 23, 1959): 215–16.
56 Willard Bascom and Roger Revelle, "Free-Diving: A New Exploratory Tool," *American Scientist* 41 (October 1953): 624.
57 Clark, "Mating of Groupers," 23.
58 Clark, *Lady and the Sharks*, 71.
59 Ibid., 11.
60 Ibid., 90–91.
61 Ibid., 95.
62 Eugenie Clark, "Instrumental Conditioning of Lemon Sharks," *Science* 130 (July 24, 1959): 217–18.
63 Coles Phinizy, "Lovely Lady with a Very Fishy Reputation," *Sports Illustrated* 23 (October 4, 1965): 50.
64 William M. Stephens, "The Lady and the Sharks," *Saturday Evening Post* 232 (July 4, 1959): 52–53.
65 Clark, *Lady and the Sharks*, 106.
66 Stewart Springer and Perry W. Gilbert, "Anti-Shark Measures," in *Sharks and Survival*, ed. Perry Gilbert (Boston: Heath, 1963), 465.
67 Perry W. Gilbert, Leonard P. Schultz, and Stewart Springer, "Shark Attacks of 1959," *Science* 132 (August 5, 1960): 323–26.
68 Eugenie Clark, "Shark Repellent Effect of the Red Sea Moses Sole," in *Shark Repellents from the Sea: New Perspectives*, ed. Bernard J. Zahuranec, AAAS Selected Symposium 83 (Boulder, Colo.: Westview Press, 1983), 135–50.
69 Clark, *Lady and the Sharks*, 109. Also see Joan Arehart-Treichel, "Demystifying the Shark: Some Things 'Jaws' Didn't Tell You," *Science News* 110 (September 4, 1976): 155–57.
70 Eugenie Clark, "Into the Lairs of 'Sleeping' Sharks," *National Geographic* 147 (April 1975): 570–84.
71 Amitai Etzioni, "From 'Jaws': A Lovable Scientist," *Science* 191 (23 January 1976): 247.
72 Eugenie Clark, "Horrors of the Deep," in *Oceans: Our Continuing Frontier*. Newspaper articles for the Firth Course by Newspaper (Del Mar, Calif.: Publisher's Inc., 1976), 14.
73 Eugenie Clark, "Sharks: Magnificent and Misunderstood," *National Geographic* 160 (August 1981): 138–87.

74 Eugenie Clark, introduction to *Sharks!* by Downs Matthews (Avenel, N.J.: Wings Books, 1996), 9.

75 Arehart-Treichel, "Demystifying the Shark," 155.

76 Ania Savage, "For Shark Anglers, It Was a Whale of a Summer," *New York Times*, September 28, 1975, 73; Ania Savage, "'Jaws' Spurs Shark Fishing," *New York Times*, October 12, 1975, 150.

77 Clark, *Lady and the Sharks*, 121–22.

78 Jack Viets, "Skin Diving for Mothers-to-Be," *San Francisco Examiner* (February 10, 1961), 12.

79 Clark, *Lady and the Sharks*, 123.

80 Clark, *Lady with a Spear*, 43.

81 Ibid., 43–44.

82 Shirley Goodstone, "At Home in the Ocean," *Doctor's Wife* (March–April, 1961), 25–27. The caption to the family portrait printed in this article reads "A Konstantinu family portrait. *Doctor Ilias*, (emphasis mine) Tak, Iris, Hera, Nicholas and Eugenie." This magazine seems to be appropriately named.

83 Phinizy, "Lovely Lady," 47.

84 Clark also notes that she couldn't "really complain about discrimination of women in the field of science. . . . I think it took harder work, perhaps, to get started, to prove myself in what was once mainly a man's field. But once over that hurdle, I always seemed to get more credit for having courage to do the same things as a man" (Clark acceptance speech for the Society of Woman Geographers Gold Medal, 78).

85 Chafe, *Paradox of Change*, 194–213.

86 A similar argument is made by Louise Newman's analysis of Margaret Mead. "Coming of Age, but Not in Samoa: Reflections on Margaret Mead's Legacy for Western Liberal Feminism," *American Quarterly* 48 (June 1996): 233–70.

87 In a 1994 interview, Clark said that "it amused me that when I did do some of the things (e.g., diving in caves with 'sleeping' sharks) considered 'macho male accomplishments' that I was given more credit than males for doing the same thing they did. It helped to balance some of the prejudices against females." Balon, "Life and Work of Eugenie Clark," 122.

CHAPTER 6: TECHNOPHILIA AND TECHNOPHOBIA IN THE OCEANIC COMMONS

1 Leo Marx, *The Machine in the Garden: Technology and the Pastoral Ideal in America* (London: Oxford University Press, 1965).

2 David E. Nye, *Narratives and Spaces: Technology and the Construction of American Culture* (New York: Columbia University Press, 1997).

3 Paul Boyer, *By the Bomb's Early Light: American Thought and Culture at the Dawn of the Atomic Age* (Chapel Hill: University of North Carolina Press, 1994).

4 Alan Nadel, *Containment Culture: American Narratives, Postmodernism, and the Atomic Age* (Durham, N.C.: Duke University Press, 1995).

5 Gorham Munson, "Adventure Writing in Our Time," *College English* 16, no. 3 (December 1954): 153.

6 Rob Wilson, *Reimagining the American Pacific: From South Pacific to Bamboo Ridge and Beyond* (Durham, N.C.: Duke University Press, 2000).

7 J. R. Eyerman, "Guam: U.S. Makes Little Island into Mighty Base," *Life* 19, no. 1 (July 2, 1945): 63.

8 John Dos Passos, "The Atolls: Outfield of Pacific War, Yesterday's Island Battlefields Are Stations on Road to Japan," *Life* 18, no. 11 (March 12, 1945): 97.

9 Warren H. Atherton, past national commander, American Legion, "Oceans of Opportunity," *American Magazine* 139, no. 4 (April 19, 1945): 36.

10 Ibid., 142.

11 Hal Friedman, *Creating an American Lake: United States Imperialism and Strategic Security in the Pacific Basin, 1945–1947* (Westport, Conn.: Greenwood Press, 2001).

12 Eugene Burdick, "Journey across the Pacific," *Holiday* 28 (October 1960): 105.

13 Laura Thompson, "The Basic Conservation Problem," *Scientific Monthly* 68 (February 1949): 129–31. Gary Kroll, "The Pacific Science Board in Micronesia: Preparation and Preservation of a New Frontier Territory," *Minerva* 41, no. 1 (2003): 25–46.

14 Jonathan M. Weisgall, *Operation Crossroads: The Atomic Tests at Bikini Atoll* (Annapolis: Naval Institute Press, 1994); Chandra Mukerji, *A Fragile Power: Scientists and the State* (Princeton, N.J.: Princeton University Press, 1989).

15 "Bikini: Breath-Holding before a Blast—Could it Split the Earth?" *Newsweek* 28, no. 1 (July 1, 1946): 20.

16 "Atomic Age: The Goodness of Man," *Time* 47, no. 13 (April 1, 1946): 28.

17 The tension described here also manifests itself in the ironic naming of the bikini bathing suit, where nuclear energy and liberation is matched with sexual energy and liberation. See John Stilgoe, *Alongshore* (New Haven, Conn.: Yale University Press, 1994).

18 Arnold Jacoby, *Señor Kon-Tiki: The Life and Adventures of Thor Heyerdahl* (New York: Rand McNally, 1967), 281.

19 Ibid., 39.

20 Ibid., 45.

21 Thor Heyerdahl, *Fatu-Hiva: Back to Nature* (New York: Doubleday, 1975), 5.

22 Thor Heyerdahl, "Turning Back Time in the South Seas," *National Geographic Magazine* 97 (January 1941): 109.

23 Heyerdahl, *Fatu-Hiva*, 24.

24 Fatu-Hiva was part of the French Society Islands. The colonial government, however, prohibited European occupation of the island. Thor and Liv had to secure special passports.

25 Robert Ascher, "Experimental Archeology," *American Anthropologist*, n.s., 63 (August 1961): 793–816.

26 Thor Heyerdahl, *Kon-Tiki: Across the Pacific by Raft* (Chicago: Rand McNally, 1950), 30.

27 Ibid., 53–54.

28 "Raft Party Begins Cross-Pacific Trip," *New York Times*, April 29, 1947, 39.

29 Heyerdahl, *Kon-Tiki*, 79.

30 "Six Men on a Raft," *New York Times*, August 12, 1947, 22:2. Also see Waldemar Kaempffert, "Kon-Tiki Raft Plays a Practical Part in an Old Argument about the Origin of Civilization," "Science in Review" *New York Times*, August, 17, 1947, 9; and

Lawrence E. Davies, "Kon-Tiki Raft Drift Gave 'Proof' of Leader's Theory on Polynesians," *New York Times,* September 30, 1947, 33:3, for similar discussions.

31 Thor Heyerdahl, "Six Men on a Raft: A Tale for Statesmen," *New York Times,* October 12, 1947, sec. VI, 18.

32 Heyerdahl, "Six Men on a Raft," 62.

33 "There Must be Something about the Sea . . . ," *New York Times Book Review,* September, 30, 1951, 19.

34 James A. Michener, "4,100 Miles on a Raft," *Saturday Review of Literature* 33 (September 23, 1950): 12. James Michener rose to great fame with his *Tales of the South Pacific* (1946) and *Return to Paradise* (1950).

35 Michener, "4,100 Miles on a Raft," 13.

36 Jacoby, *Señor Kon-Tiki,* 279.

37 Susan Heller Anderson, "A Social Institution Called Trader Vic's," *New York Times,* October 17, 1979, C3.

38 Trader Vic, *Trader Vic's Book of Food and Drink* (New York: Doubleday, 1946), 17–22.

39 David Farber and Beth Bailey, "The Fighting Man as Tourist: The Politics of Tourist Culture in Hawaii during World War II," *Pacific Historical Review* 65 (November 1996): 643. John S. Whitehead, "Noncontinguous Wests: Alaska and Hawai'i," in *Many Wests: Place, Culture and Regional Identity,* ed. David M. Wrobel and Michael C. Steiner (Lawrence: University Press of Kansas, 1997), 314–41.

40 Farber and Bailey, "Fighting Man as Tourist," 659.

41 David M. Wrobel, "Introduction: Tourists, Tourism and the Toured Upon," 1–34, and Patricia Nelson Limerick, "Seeing and Being Seen: Tourism in the American West," in *Seeing and Being Seen: Tourism in the American West,* ed. David M. Wrobel and Patrick T. Long (Lawrence: University Press of Kansas, 2001), 39–58.

42 Ben Finney and James Houston, *Surfing: A History of the Ancient Hawaiian Sport* (San Francisco: Pomegranate Artbooks, 1996); Michael Nevin Willard, "Duke Kahanamokiu's Body: Biography of Hawai'i," in *Sports Matters: Race, Recreation, and Culture,* ed. John Bloom and Michael Nevin Willard (New York: New York University Press, 2002), 13–38; Arthur C. Verge, "George Freeth: King of the Surfers and California's Forgotten Hero," *California History* 80 (2001): 82–105.

43 Douglas Booth, "Ambiguities in Pleasure and Discipline: The Development of Competitive Surfing," *Journal of Sports History* 22 (fall 1995): 192; Jon Krakauer, "Gidget Has Grown Up, but Surfing Is Still a 'Totally Happening' Sport," *Smithsonian* 20 (June 1989): 106–19.

44 Douglas Booth, "Surfing Films and Videos: Adolescent Fun, Alternative Lifestyle, Adventure Industry," *Journal of Sports History* 23 (fall 1996): 323.

45 Quoted in Booth, "Ambiguities in Pleasure and Discipline," 193, 196.

46 Jacoby, *Señor Kon-Tiki,* 281.

47 Munson, "Adventure Writing in Our Time," 55.

48 Thor Heyerdahl, *The Ra Expeditions* (Garden City, N.Y.: Doubleday, 1971), 335.

49 Thor Heyerdahl, "How Vulnerable Is the Ocean?", in *Who Speaks for Earth?,* ed. Maurice F. Strong (New York: W. W. Norton, 1973), 46–49.

50 Ibid., 61–63.

51 James Dugan, "Portrait of Homo Aquaticus," *New York Times,* April 21, 1963, SM20.

52 Capt. Jacques-Yves Cousteau, "Fish Men Explore a New World Undersea," *National Geographic Magazine* 102 (October 1952): 431.

53 Wendell P. Bradley, "French Co-Inventor of Aqualung Here to Inspect Navy's Diving School," *Washington Post* (November 29, 1953), M25.

54 "Underwater Wonders: A French Diver Explores the Haunts of Sea Monsters," *Life* (November 27, 1950), 119–25, 431–32.

55 Capt. Jacques-Yves Cousteau, "Fish Men Explore a New World Undersea," 431–472.

56 "Ancient Cargo Ship Is Being Salvaged," *New York Times,* March 1, 1953, 15.

57 Capt. Jacques-Yves Cousteau, "Fish Men Discover a 2,200-Year-Old Greek Ship," *National Geographic Magazine* 105 (January 1954): 3–34.

58 The excavation was also subject of a January 1954 Omnibus episode called *Underseas Archeology.* Lawrence Laurent, "That 'Omnibus' Audience Likes Its Territory New," *Washington Post* (January 17, 1954), L1.

59 Capt. Jacques-Yves Cousteau, "Calypso Explores for Underwater Oil," *National Geographic Magazine* 108 (August 1955): 157–84.

60 "New Oil Exploration," *New York Times,* October 31, 1953, 21.

61 Jacques-Yves Cousteau, "Exploring Davy Jones's Locker with Calypso," *National Geographic Magazine* 109 (February 1956): 159.

62 "Deep-Water Diving Turtle Passes Tests," *Washington Post* (October 16, 1959), D4.

63 Jacques Cousteau, "Inflatable Ship Opens Era of Airborne Undersea Expeditions," *National Geographic Magazine* 120 (July 1961): 142–48.

64 Dugan, "Portrait of Homo Aquaticus," SM20.

65 Jacques Cousteau, "Ocean-Bottom Homes for Skin Divers," *Popular Mechanics* 120 (July 1963): 98.

66 "Cousteau Planning Report in Cinerama," *New York Times,* November 26, 1963, 53.

67 "Oceanography: Home in the Deep," *Time* 82 (July 26, 1963): 40.

68 Jacques Cousteau, "At Home in the Sea," *National Geographic Magazine* 125 (April 1964), 472–73. Ramon Geremia, "Beauty, Bounty Lie under the Sea," *Washington Post* (November 7, 1964), D2.

69 Geremia, "Beauty, Bounty Lie Under Sea," D2.

70 Jacques Cousteau, "Working for Weeks on the Sea Floor," *National Geographic Magazine* 126 (April 1966): 533.

71 Jack Gaver, "Undersea Pictures Are Hard to Take," *Washington Post* (April 15, 1966), D12.

72 Cousteau, "Working for Weeks on the Sea Floor," 536.

73 "Cousteau 'Island' Wrecked by Fire," *New York Times,* February 21, 1965, 14; "Trial of Diving Bell Pleases Cousteau," *Washington Post* (August 30, 1965), E4.

74 William Smith, "Oceanography: The Profit Potential Is as Big as the Sea," *New York Times,* January 31, 1968, 8, 12; Monty Hoyt, "Ocean Mining Aims at 'Dry-Land' Technique," *Christian Science Monitor* (May 28, 1970).

75 *Our Nation and the Sea: A Plan for National Action,* Report of the Commission on Marine Science, Engineering and Resources (Washington, D.C.: U.S. Government Printing Office, 1969), 72.

76 Leonard Sloane, "Newest Rigs for Oil Drilling Will Be Submerged in the Oceans," *New York Times,* June 9, 1968, F11.

77 George F. Bond, *Papa Topside: The Sealab Chronicles of Capt. George F. Bond, USN* (Annapolis: Naval Institute Press, 1993), 53.

78 Ibid., 81–82.

79 "Aquanauts Begin Sealab Tests; Carpenter Speaks to Gemini 5," *New York Times*, August 30, 1965.

80 Joseph N. Bell, "Chilly, Wet 'Walk'; Window Under Water," *Christian Science Monitor* (October 27, 1965).

81 Jack Gould, "TV Review," *New York Times*, March 19, 1965.

82 Robert Cowen, "U.S. Aquanauts 'Think Deep'; Major Development Needed," *Christian Science Monitor* (November 8, 1968).

83 Richard Lyons, "Oceanographers Scout Sea Floor for Sites to Put Manned Centers," *New York Times*, August 1, 1968.

84 Kenneth G. Slocum, "Exploiting the Deep: Tapping Seas Wealth Sparks Interest, but Payoff is Remote," *Wall Street Journal*, September 16, 1968, 1, 24.

85 Gloria Emerson, "Cousteau to Take TV Films of Seas," *New York Times*, March 5, 1967, 24.

86 Walter Sullivan, "TV: Jacques Cousteau Visits World of the Sharks," *New York Times*, January 9, 1968, 87.

87 Howard Thompson, "Series on Marine Life Begins on NBC," *New York Times*, June 7, 1973, 91.

88 Sandra Blakeslee, "Cousteau and Satellites Check Health of the Antarctic," *New York Times*, February 9, 1973, 69.

89 Gerald Weales, "Cousteau? Hippo, Hippo, Horray!" *New York Times*, March 18, 1973, 145.

90 Jacques Cousteau, *The Ocean World* (New York: Harry N. Abrams, 1972), 1.

91 Jacques Cousteau, "Our Oceans are Dying," *New York Times*, November 14, 1971, E13.

92 Nancy Hicks, "Cousteau's Philosophy of the Sea Helps Get Him Another Medal," *New York Times*, October 25, 1970, 54; "The Dying Oceans," *Time* 96 (September 28, 1970): 64.

93 Peter Taylor, "Technocratic Optimism, H. T. Odum and the Partial Transformation of Ecological Metaphor after World War II," *Journal of the History of Biology* 21 (1988): 213–44.

94 Andrew Kirk, "Appropriating Technology: Alternative Technology, the Whole Earth Catalog and Counterculture Environmental Politics," *Environmental History* 6 (July 2001): 374–94.

95 Jacques Cousteau, "The Perils and Potentials of a Watery Planet," *Saturday Review World* (August 24, 1974), 41–44, 122.

96 Axel Madsen, *Cousteau: An Unauthorized Biography* (New York: Beaufort Books, 1986), 172.

97 "A Friend of Yours is . . . ," *Calypso Log* 1, Newsletter of the Cousteau Society (July 1974).

98 "International Report," Philippe Cousteau (editor), *Calypso Log* 1, Newsletter of the Cousteau Society (July 1974).

99 "Oasis in Space," *Calypso Log* 2, Newsletter of the Cousteau Society (November–December 1975).

100 Madsen, *Cousteau*, 186. Also see Cousteau Society information literature, "Ecotechnie." Interview with Jacques Cousteau, "Consumer Society Is the Enemy," *New Perspectives Quarterly* 16 (spring 1999): 17.

101 David Nye, *America as Second Creation: Technology and Narratives of New Beginnings* (Cambridge, Mass.: MIT Press, 2003).

1 "Nest of Naiads: Female Scientists Live Undersea for Project Tektite," *Life* 69 (July 17, 1970): 30–31; photograph in J. G. VanDerwalker, "Science's Window on the Sea," *National Geographic* 140 (August 1971): 256; Sylvia Earle, "All-Girl Team Tests the Habitat," *National Geographic* 140 (August 1971): 296.
2 *Our Nation and the Sea: A Plan for National Action,* Report of the Commission on Marine Science, Engineering and Resources. (Washington, D.C.: U.S. Government Printing Office, 1969), 1.
3 Dorinda Dallmeyer, ed., *Values at Sea: Ethics for the Marine Environment* (Athens: University of Georgia Press, 2003).
4 This is not completely true. Depending on how postmodern we want to be, it is equally valid to consider the American West a cultural construct. Although I am sensitive to this important argument, I must ignore this and use some sort of language that refers to space. See Martin Lewis and Karen Wigen, *The Myth of Continents: A Critique of Metageography* (Berkeley: University of California Press, 1997).
5 William Cronon, "The Trouble with Wilderness, or, Getting Back to the Wrong Nature," in *Uncommon Ground: Toward Reinventing Nature,* ed. William Cronon (New York: W. W. Norton, 1995), 69–90.

Selected Bibliography

MANUSCRIPT SOURCE LOCATIONS

Andrews Papers/DM
American Museum of Natural History, Department of Mammalogy, Manhattan, N.Y.

AMNH Special Collections
American Museum of Natural History Library, Special Collections, Manhattan, N.Y.
Personnel Files
Exhibit Files
Expedition Files

Beebe Papers/NYZS, DTR
New York Zoological Society Library, Bronx Zoo, New York
Office of the Director and General Associate, Group 58
Office of the General Associate, Group 59
General Records, Group 60
Oceanographic Expeditions—Vessels *Noma* and *Arcturus,* Group 65
Oceanographic Expeditions—Vessel *Zaca,* 1936–1938, Group 68

Beebe Papers/PUL
Firestone Library, Rare Books and Manuscripts, Princeton University, Princeton, N.J.
William Beebe Collection, Manuscript C0661

Carson Papers
Beinecke Rare Book and Manuscript Archive, Yale Collection of American Literature, Yale University, New Haven, Conn.
Rachel Carson Manuscript Collection

Lear Carson Papers
Rare Books and Manuscript Collection, Connecticut College, New London, Conn.

Murphy Papers
American Philosophical Society Library, Manuscript Collections, Philadelphia, Pa.
Robert Cushman Murphy papers, B: M957

PSB Papers
National Academy of Science Archive, Washington, D.C.
National Research Council—Pacific Science Board

PUBLISHED SOURCES

Adams, Carol J. *The Sexual Politics of Meat: A Feminist-Vegetarian Critical Theory.* New York: Continuum, 1990.

Allen, Garland. *Life Sciences in the Twentieth Century.* Cambridge: University of Cambridge Press, 1978.

———. "Old Wine in New Bottles: From Eugenics to Population Control in the Work of Raymond Pearl." In *The Expansion of American Biology,* edited by Keith Benson, Jane Maienschein, and Ronald Rainger, 231–61. New Brunswick, N.J.: Rutgers University Press, 1991.

Allen, Joel. "The North Atlantic Right Whale and Its Near Allies." *Bulletin of the American Museum of Natural History* 24 (1908): 277–329.

Andrews, Roy Chapman. *Across Mongolian Plains: A Naturalist's Account of China's "Great Northwest."* New York: Blue Ribbon Books, 1921.

———. *This Business of Exploring.* New York: G. P. Putnam's Sons, 1935.

———. "Explorers and Their Work." *Saturday Evening Post* 204 (August 22, 1931), 6–7, 84–85.

———. "Monographs of the Pacific Cetacean, I—The California Gray Whale. Its History, Habits, External Anatomy, Osteology and Relationship." *Memoirs of the American Museum of Natural History,* n.s., vol. 1, pt. 5 (1915): 229–90.

———. "Shore-Whaling: A World Industry." *National Geographic Magazine* 22 (May 1911): 411–42.

———. *Under a Lucky Star: A Lifetime of Adventure.* New York: Viking Press, 1944.

———. "Whale-Hunting as it is Now Done." *World's Work* 17 (December 1908): 11031–47.

———. *Whale Hunting with Gun and Camera.* New York: D. Appleton, 1916.

———. "What Shore-Whaling Is Doing for Science." *Nature* 88 (December 28, 1911): 280–82.

Anderson, Karen. *Wartime Women: Sex Roles, Family Relations and the Status of Women during World War II.* Westport, Conn.: Greenwood Press, 1981.

Angell, Roger. "World of Women, USA: Lady with a Spear." *Holiday* 17 (January 1955): 48–51.

Anthony, H. E. "Glimpses into the Hall of Ocean Life." *Natural History* 33 (July–August 1933): 365+.

Arehart-Treichel, Joan. "Demystifying the Shark: Some Things 'Jaws' Didn't Tell You." *Science News* 110 (September 4, 1976): 155–57.

Aron, Stephen. "Lessons in Conquest: Towards a Greater Western History." *Pacific Historical Review* 63 (May 1994): 125–48.

Ascher, Robert. "Experimental Archeology." *American Anthropologist,* n.s., 63 (August 1961): 793–816.

Balon, Eugene. "The Life and Work of Eugenie Clark: Devoted to Diving and Science." *Environmental Biology of Fishes* 41 (1994): 89–114.

Bardach, John. *Harvest of the Sea.* New York: Harper and Row, 1968.

Bascom, Willard, and Roger Revelle. "Free-Diving: A New Exploratory Tool." *American Scientist* 41 (October 1953): 624–27.

Bates, Marston. *The Nature of Natural History.* London: Chapman & Hall, 1951.

Beebe, William. *The Arcturus Adventure: An Account of the New York Zoological Society's First Oceanographic Expedition.* New York: Harper and Row, 1981.

———. *Beneath Tropical Seas.* New York: Blue Ribbon Books, 1928.

———. *The Book of Naturalists: An Anthology of the Best Natural History.* New York: Alfred A. Knopf, 1944.

———. "Descent into Perpetual Night." *New York Times Magazine,* October 9, 1932, 1+.

———. "Down into Davy Jones's Locker." *New York Times Magazine,* July 13, 1930, 1+.

———. "The Evolution and Destruction of Life." *Zoological Society Bulletin* 21 (May 1918): 1622–24.

———. "A Half Mile Down." *National Geographic Magazine* 66 (December 1934): 661–75.

———. "The Jelly-Fish and Equal Suffrage." *Atlantic Monthly* 114 (July 1914): 36–47.

———. "Jungle Night." *Atlantic Monthly* 120 (July 1917): 69–79.

———. *Jungle Peace.* New York: H. Holt, 1918.

———. "Lord Dunsany and 'Don Rodriguez.'" *New York Times Review of Books and Magazine,* October 1, 1922, 3.

———. "In Pursuit of an Elusive Sea." *New York Times Magazine,* April 19, 1925, 1+.

———. "A Quarter Mile Down in the Open Sea." *Bulletin: New York Zoological Society* 33 (November–December 1930): 200–231.

———. Review of *Titans of the Deep,* film by Otis Barton. In *Science,* April 7, 1939, 319.

———. Review of *Under the Sea-Wind,* by Rachel Carson. In *Saturday Review of Literature,* December 27, 1941, 5.

———. "Round Trip to Davy Jones' Locker." *National Geographic Magazine* 59 (June 1931): 635–78.

———. "Seventy-Four: An Island of Water." *Atlantic Monthly* 137 (January 1926): 37–45.

———. "Thoughts on Diving." *Harper's Monthly* 166 (April 1933): 582–84.

———. "A Wilderness Laboratory." *Atlantic Monthly* 119 (May 1917): 628–38.

———. "A Yard of Jungle." *Atlantic Monthly* 117 (January 1916): 40–47.

Beebe, William and Mary Blair Beebe. "A Naturalist in the Tropics." *Harper's Monthly* 118 (March 1909): 590–600.

Beebe, William, John Tee-Van, Gloria Hollister, Jocelyn Crane, and Otis Barton. *Half-Mile Down.* New York: Harcourt, Brace, 1934.

Bender, Bert. *Sea-Brothers: The Tradition of American Sea Fiction from Moby Dick to the Present.* Philadelphia: University of Pennsylvania, 1988.

Benson, Keith R. "From Museum Research to Laboratory Research: The Transformation of Natural History into Academic Biology." In *The American Development of Biology,* edited by Ronald Rainger, Keth Benson, and Jane Maienschein, 49–86. New Brunswick, N.J.: Rutgers University Press, 1991.

Betson, Henry. "The Outermost House." In *Especially Maine: The Natural World of Henry Betson from Cape Cod to the St. Lawrence,* edited by Elizabeth Coatsworth. Brattleboro, Vt.: Stephen Greene Press, 1970.

Berrill, Michael. *The Plundered Seas: Can the World's Fish be Saved?* San Francisco: Sierra Club Books, 1997.

Bigelow, Henry. *Oceanography: Its Scope, Problems, and Economic Importance.* Boston: Houghton Mifflin, 1931.

Biurnbaum, Martin. "Vanishing Eden." *Natural History* 49 (March 1942): 161–71.

Blake, Kevin S. "Zane Grey and Images of the American West." *Geographical Review* 85, no. 2 (April 1995): 202–16

Blanton, Casey. *Travel Writing: The Self and the World.* New York: Simon & Schuster, 1997.

Bloom, Lisa. *Gender on Ice: American Ideologies of Polar Expedition.* Minneapolis: University of Minnesota Press, 1993.

Boesen, Victor Boesen. "Adventures of a Shark Diver." *Saturday Evening Post* 224 (July 14, 1951): 18+.

Bond, George F. *Papa Topside: The Sealab Chronicles of Capt. George F. Bond, USN.* Annapolis: Naval Institute Press, 1993.

Bonta, Marcia Myers. *Women in the Field: America's Pioneering Women Naturalists.* College Station: Texas A&M University Press, 1991.

Booth, Douglas. "Ambiguities in Pleasure and Discipline: The Development of Competitive Surfing." *Journal of Sports History* 22 (fall 1995): 192.

———. "Surfing Films and Videos: Adolescent Fun, Alternative Lifestyle, Adventure Industry." *Journal of Sports History* 23 (fall 1996): 323.

Borth, C. Carl. "A Modern Marco Polo." *Bermudian,* November 1931, 12+.

Bowler, Peter. *The Environmental Sciences.* New York: W. W. Norton, 1992.

Boyer, Paul. *By the Bomb's Early Light: American Thought and Culture at the Dawn of the Atomic Age.* Chapel Hill: University of North Carolina Press, 1994.

Bradley, David. *No Place to Hide.* Boston: Little, Brown, 1948.

Breder, C. M., and Eugenie Clark, "A Contribution to the Visceral Anatomy, Development, and Relationships of the Plectognathi." *Bulletin of the American Museum of Natural History* 88, article 5 (1947): 287–320.

Breit, Harvey. Review of *Sea Around Us* by Rachel Carson. In *Atlantic* 188 (August 1951): 84.

Bridges, William. *Gathering of Animals: An Unconventional History of the New York Zoological Society.* New York: Harper & Row, 1974.

Brinkley, Alan. *Voices of Protest: Huey Long, Father Coughlin and the Great Depression.* New York: Vintage Books, 1982.

Brooks, Paul. *The House of Life: Rachel Carson at Work, with Selections from Her Writings.* Boston: Houghton Mifflin, 1972.

Buell, Lawrence. *The Environmental Imagination: Thoreau, Nature Writing, and the Formation of American Culture.* Cambridge, Mass.: Harvard University Press, 1995.

Camerini, Jane. "Wallace in the Field." In *Science in the Field,* edited by Henrika Kuklick and Robert E. Kohler. *Osiris,* 2nd ser., 11 (1996): 44–65.

"Career Women." *Cornet* 33 (April 1953): 115–28.

Carson, Rachel. "The Bat Knew it First." *Collier's,* November 18, 1944.

———. "The Birth of an Island." *Yale Review* 40 (September 1950): 112–26.

———. *The Edge of the Sea.* New York: Houghton Mifflin, 1955.

———. *Guarding Our Wildlife Resources. Conservation in Action,* no. 5. Washington, D.C.: U.S. Fish and Wildlife Service, Government Printing Office, 1947.

———. *The Sea Around Us.* New York: Oxford University Press, 1951.

———. *Under the Sea-Wind: A Naturalist's Picture of Ocean Life.* 2nd ed. New York: Oxford University Press, 1952.

———. "Undersea." *Atlantic Monthly* 160 (September 1937): 322–25.
Carnes, Mark C. *Secret Ritual and Manhood in Victorian America.* New Haven, Conn.: Yale University Press, 1989.
Catton, Ted. *Inhabited Wilderness: Indians, Eskimos, and National Parks in Alaska.* Albuquerque: University of New Mexico Press, 1997.
Chafe, William. *The Paradox of Change: American Women in the 20th Century.* New York: Oxford University Press, 1991.
Chandler, Alfred. *Strategy and Structure: Chapters in the History of the Industrial Enterprise.* Cambridge: Harvard University Press, 1962.
Clark, Eugenie."Field Trip to the South Seas." *Natural History* 60 (1951): 8–16.
———. "Functional Hermaphroditism and Self-Fertilization in a Serranid Fish." *Science* 129 (January 23, 1959): 215–16.
———. "Horrors of the Deep." In *Oceans: Our Continuing Frontier,* 13–15. San Diego: Publisher's Inc., 1976.
———. "Instrumental Conditioning of Lemon Sharks." *Science* 130 (July 24, 1959): 217–18.
———. "Into the Lairs of 'Sleeping' Sharks." *National Geographic* 147 (April 1975): 570–84.
———. Introduction to *Sharks!* by Downs Matthews. Avenel, N.J.: Wings Books, 1996.
———. *Lady and the Sharks.* Sarasota, Fla.: Mote Marine Laboratory, 1969.
———. *Lady with a Spear.* New York: Harper & Brothers, 1953.
———. "The Lost Quarry." *Natural History* 61 (1952): 258–63.
———. "Mating of Groupers." *Natural History* 74 (June 1965): 22–25.
———. "A Method for Artificial Insemination in Viviparous Fishes." *Science* 112 (December 15, 1950): 722–23.
———. "Reef Fish Studies in the South Pacific." *Scientific Investigations in Micronesia #1.* Washington, D.C.: National Research Council, 1949.
———. Review of *Diving to Adventure* by Hans Haas. *Natural History* 61 (January 1952): 4.
———. "A Scientific Journey to the Red Sea, Part 1." *Natural History* 61 (1952): 344–49.
———. "A Scientific Journey to the Red Sea, Part 2," *Natural History* 61 (1952): 414–19
———. "Shark Repellent Effect of the Red Sea Moses Sole." In *Shark Repellents from the Sea: New Perspectives,* edited by Bernard J. Zahuranec, 135–50. AAAS Selected Symposium 83. Boulder, Colo.: Westview Press, 1983.
———. "Sharks: Magnificent and Misunderstood." *National Geographic* 160 (August 1981): 138–87.
———. "Siakong, Spear-Fisherman Pre-eminent." *Natural History* 62 (1953): 227–34.
———. "Wonders of the Deep, by Eugenie Clark." *Cairo Calling* 847, pt. 2 (June 9, 1951): 12–13.
Clark, Eugenie, and Robert Kamrin "The Role of the Pelvic Fins in the Copulatory Act of Certain Poeciliid Fishes." *American Museum Novitates* 1509 (May 7, 1951): 1–13.
Clark, Eugenie, et al. "Social Behavior of Caribbean Tilefish." *Underwater Naturalist* 18 (1989): 20–23.
Claudy, C. H. "The Whale as a Food Factor." *Scientific American* 118 (March 9, 1918): 208–9.
Clause, Bonnie Tocher. "The Wistar Rat as a Right Choice: Establishing Mammalian Standards and the Ideal of a Standardized Mammal." *Journal of the History of Biology* 26 (1993): 329–50.

Coats, Peter. "Amchitka, Alaska: Toward the Bio-Biography of an Island." *Environmental History* 1 (October 1996): 20–45.

———. *This Great and Wide Sea.* Chapel Hill: University of North Carolina Press, 1947.

———. "Peru's Wealth-Producing Birds." *National Geographic Magazine* 37 (June 1920): 537–66.

Conrad, Joseph. "Ocean Travel." In *Last Essays,* edited by Richard Curle, 35–38. Garden City, N.Y.: Doubleday, Page, 1926.

Cooter, Roger and Stephen Pumfrey. "Separate Spheres and Public Places: Reflexions on the History of Science Popularization and Science in Popular Culture." *History of Science* 32 (1994): 237–67.

Corn, Joseph J. *The Winged Gospel: America's Romance with Aviation, 1900–1950.* New York: Oxford University Press, 1983.

Cousteau, Capt. Jacques-Yves. "At Home in the Sea." *National Geographic Magazine* 125 (April 1964): 465–507.

———. "Calypso Explores for Underwater Oil." *National Geographic Magazine* 108 (August 1955): 155–84..

———. "Exploring Davy Jones's Locker with Calypso." *National Geographic Magazine* 109 (February 1956): 149–61.

———. "Fish Men Discover a 2,200-Year-Old Greek Ship." *National Geographic Magazine* 105 (January 1954): 1–36.

———. "Fish Men Explore a New World Undersea." *National Geographic Magazine* 102 (October 1952): 431–72.

———. "Inflatable Ship Opens Era of Airborne Undersea Expeditions." *National Geographic Magazine* 120 (July 1961): 142–48.

———. *The Ocean World.* New York: Harry N. Abrams, 1972.

———. "Working for Weeks on the Sea Floor." *National Geographic Magazine* 129 (April 1966): 498–537.

Crawford, Fred D. *H. M. Tomlinson.* Boston: Twayne Publishers, 1981.

Cronon, William. *Nature's Metropolis: Chicago and the Great West.* New York: W. W. Norton, 1991.

———. "The Trouble with Wilderness; or, Getting Back to the Wrong Nature." In *Uncommon Ground: Toward Reinventing Nature,* edited by William Cronon, 69–90. New York: W. W. Norton, 1995.

Cushman, Gregory T. "'The Most Valuable Birds in the World': International Conservation Science and the Revival of Peru's Guano Industry, 1909–1965." *Environmental History* 10, no. 3 (July 2005): 477–509.

Cutright, Paul. *Theodore Roosevelt: The Making of a Conservationist.* Urbana: University of Chicago Press, 1985.

Dallmeyer, Dorinda, ed. *Values at Sea: Ethics for the Marine Environment.* Athens: University of Georgia Press, 2003.

Darby, Hugh. Review of *A Half-Mile Down* by William Beebe et al. *Nation* 139 (December 12, 1934): 687.

Desmond, Adrian. "Artisan Resistance and Evolution in Britain, 1819–1848." *Osiris,* 2nd ser., 3 (1987): 77–110.

de Latil, Pierre, and Mary Thayer Muller. "Wonderland of Deep Sea Hunters." *Coronet* 32 (September 1952): 38–42.

Douglas, William O. "Environmental Problems of the Oceans: The Need for International Controls." *Environmental Law* 2 (spring 1971): 149–66.
Dorsey, Kurkpatrick. *The Dawn of Conservation Diplomacy: U.S.-Canadian Wildlife Protection Treaties in the Progressive Era.* Seattle: University of Washington Press, 1998.
Dubbert, Joe. "Progressivism and the Masculinity Crisis." In *The American Man,* edited by Elizabeth Pleck and Joseph Pleck, 303–20. Englewood Cliffs, N.J.: Prentice-Hall, 1980.
Duffus, R. L. "'Arcturus,' Whither Away?" *New York Times Book Review,* May 23, 1926, 1.
Duncan, William. "Life a Mile Down in the Sea Described by Dr. William Beebe." (Philadelphia, Pa.) *Public Ledger,* January 18, 1933.
Earle, Sylvia. *Sea Change: A Message of the Oceans.* New York: G. P. Putnam's Sons, 1995.
Edmond, Rod. *Representing the South Pacific: Colonial Discourse from Cook to Gauguin.* New York: Cambridge University Press, 1997.
Egan, Ferol. *Fremont: Explorer for a Restless Nation.* Garden City, N.Y.: Doubleday, 1977.
Ellis, Richard. *The Empty Ocean: Plundering the World's Marine Life.* Washington, D.C.: Island Press/Shearwater Books, 2003.
Ellis, Richard. *Men and Whales.* 2nd ed. New York: First Lyons Press, 1999.
Entrikin, J. Nicholas. *The Betweeness of Place: Towards a Geography of Modernity.* Baltimore: Johns Hopkins University Press, 1991.
Farber, David, and Beth Bailey, "The Fighting Man as Tourist: The Politics of Tourist Culture in Hawaii during World War II." *Pacific Historical Review* 65 (November 1996): 641–60.
Farber, Paul Lawrence. *Finding Order in Nature: The Naturalist Tradition from Linnaeus to E. O. Wilson.* Baltimore: Johns Hopkins University Press, 2000.
Field, J. G.., Gotthil Hempel, and C. P. Summerhays. *Oceans 2020: Science, Trends, and the Challenge of Sustainability.* Washington, D.C.: Island Press, 2002.
Filene, Peter. "Between a Rock and a Soft Place: A Century of American Manhood." *South Atlantic Quarterly* 84 (autumn 1985): 339–55.
Filene, Peter. *Him/Her/Self: Sex Roles in Modern America.* 2nd ed. Baltimore: Johns Hopkins University Press, 1986.
Finney, Ben, and James Houston. *Surfing: A History of the Ancient Hawaiian Sport.* San Francisco: Pomegranate Artbooks, 1996.
Fox, Stephen. *John Muir and His Legacy: The American Conservation Movement.* Boston: Little, Brown, 1981.
Freeman, Martha, ed. *Always, Rachel: The Letters of Rachel Carson and Dorothy Freeman, 1952–1964.* Boston: Beacon Press, 1995.
Freud, Sigmund. *Civilization and its Discontents.* New York: W. W. Norton & Co., 1961.
Friedman, Hal. *Creating an American Lake: United States Imperialism and Strategic Security in the Pacific Basin, 1945–1947.* Westport, Conn.: Greenwood Press, 2001.
Friedman, Robert Marc. *Appropriating the Weather: Vilhelm Bjerknes and the Construction of a Modern Meteorology.* Ithaca: Cornell University Press, 1989.
Fritzell, Peter. *Nature Writing and America: Essays upon a Cultural Type.* Ames: Iowa State University Press, 1990.
Fujita, Rodney. *Heal the Ocean: Solutions for Saving our Seas.* Gabriola Island, B.C.: New Society Publishers, 2003.
Gaddis, John Lewis. *Strategies of Containment: A Critical Appraisal of Postwar American National Security Policy.* New York: Oxford University Press, 1982.
Gartner, Carol B. *Rachel Carson.* New York: Frederick Ungar Publishing, 1983.

———. "When Science Writing Becomes Literary Art." In *And No Birds Sing: Rhetorical Analyses of Rachel Carson's Silent Spring,* edited by Craig Waddell, 103–25. Carbondale: Southern Illinois University Press, 2000.

Gilbert, Perry, ed. *Sharks and Survival.* Boston: Heath, 1963.

Glover, Linda, and Sylvia Earle. *Defying Ocean's End: An Agenda for Action.* Washington, D.C.: Island Press, 2004.

Goetzmann, William H. *Army Exploration in the American West.* New Haven, Conn.: Yale University Press, 1959.

———. *Exploration and Empire: The Explorer and the Scientist in the Winning of the American West.* New York: W. W. Norton, 1966.

———. *New Lands, New Men: America and the Second Great Age of Discovery.* New York: Viking Press, 1986.

Goodstone, Shirley. "At Home in the Ocean." *Doctor's Wife,* March–April 1961, 25–27.

Gornick, Vivian. *Women in Science: Portraits from a World in Transition.* New York: Simon and Schuster, 1983.

Gottlieb, Robert. *Forcing of the Spring: The Transformation of the American Enviromental Movement.* Washington, D.C.: Island Press, 1993.

Gould, Carol Grant. *The Remarkable Life of William Beebe: Explorer and Naturalist.* Washington, D.C.: Island Press/Shearwater Books, 2004.

Greene, Mott T. "Oceanography's Double Life." *Earth Sciences History* 12 (1993): 48–53.

Grinnell, Mrs. Oliver C. "Introduction to American Big Game Fishing." In *American Big Game Fishing,* edited by Eugene V. Connett, xiv–xv. New York: Derrydale Press, 1935.

Grove, Richard. *Green Imperialism: Colonial Expansion, Tropical Island Edens, and the Origins of Environmentalism, 1600–1860.* Cambridge: Cambridge University Press, 1995.

Hagen, Joel. *An Entangled Bank: The Origins of Ecosystem Ecology.* New Brunswick, N.J.: Rutgers University Press, 1992.

Halle, Louis. *Spring in Washington.* New York: William Sloane Associates, 1947.

Haraway, Donna. *Primate Visions: Gender, Race, and Nature in the World of Modern Science.* New York: Routledge, 1989.

Harmond, Richard. "Robert Cushman Murphy and Environmental Issues on Long Island." *Long Island Historical Journal* 8 (fall 1995): 76–82.

Hartmann, Susan M. *The Home Front and Beyond: American Women in the 1940s.* Boston: Twayne Publishers, 1982.

Haynes, Roslynn. *From Faust to Strangelove: Representations of the Scientist in Western Literature.* Baltimore: Johns Hopkins University Press, 1994.

Hays, Samuel P. *Beauty, Health, and Permanence: Environmental Politics in the United States, 1955–1985.* Cambridge: Cambridge University Press, 1993.

———. *Conservation and the Gospel of Efficiency: The Progressive Conservation Movement, 1890–1920.* Cambridge: Harvard University Press, 1959.

Hellman, Geoffrey. *Bankers, Bones and Beetles: The First Century of the American Museum of Natural History.* Garden City, N.Y.: Natural History Press, 1968.

Helvarg, David. *Blue Frontier: Saving America's Living Seas.* New York: W. H. Freeman, 2001.

Herold, Daphna, and Eugenie Clark. "Monogamy, spawning and skin-shedding of the sea moth, *Eurypegasus draconis.*" *Environmental Biology of Fishes* 37 (1993): 219–36.

Hevley, Bruce. "The Heroic Science of Glacier Motion." In *Science in the Field,* edited by Henrika Kuklick and Robert Kohler. *Osiris,* 2nd ser., 11 (1996): 66–86.

Heyerdahl, Thor. *Fatu-Hiva: Back to Nature.* New York: Doubleday, 1975.
———. "How Vulnerable Is the Ocean?" In *Who Speaks for Earth?,* edited by Maurice F. Strong, 46–49. New York: W. W. Norton, 1973.
———. *The Ra Expeditions.* Garden City, N.Y.: Doubleday, 1971.
———. "Six Men on a Raft: A Tale for Statesmen." *New York Times,* October 12, 1947, sec. VI, 18.
———. "Turning Back Time in the South Seas." *National Geographic Magazine* 97 (January 1941): 109.
Hicks, Philip Marshall. "The Development of the Natural History Essay in American Literature." Ph.D. thesis. University of Pennsylvania, 1924.
Hubbs, Carl. "Reviews and Comments." *Copeia* 2 (July 16, 1935): 105.
Hubler, Richard. "Between the Devilfish and the Deep Blue Sea." *Collier's* 131 (January 24, 1953): 40–43.
Hudson, Gill. "Unfathering the Thinkable: Gender, Science and Pacifism in the 1930s." In *Science and Sensibility: Gender and Scientific Enquiry, 1780–1945,* 264–86. Cambridge, Mass.: Basil Blackwell, 1991.
Hughes, Thomas P. *American Genesis: A Century of Invention and Technological Enthusiasm.* New York: Penguin Books, 1989.
Isenberg, Andrew. *The Destruction of the Bison: An Environmental History, 1750–1920.* New York: Cambridge University Press, 2000.
Jacoby, Arnold. *Señor Kon-Tiki: The Life and Adventures of Thor Heyerdahl.* New York: Rand McNally, 1967.
Jacoby, Karl. *Crimes against Nature: Squatters, Poachers, Thieves, and the Hidden History of American Conservation.* Berkeley: University of California Press, 2001.
Jamison, Andrew, and Ron Eyerman. *Seeds of the Sixties.* Berkeley: University of California Press, 1994.
Kaempffert, Waldemar. "Into the Black Deeps of the Sea." *New York Times Magazine,* August 28, 1932, 8.
Kanigel, Robert. *The One Best Way: Frederick Winslow Taylor and the Enigma of Efficiency.* New York: Viking Press, 1997.
Kaplan, Moise N. *Big Game Fishermen's Paradise: A Complete Treatise.* Tallahassee, Fla.: Rose Printing, 1936.
Kennedy, John Michael. "Philanthropy and Science in New York City: The American Museum of Natural History." Ph.D. dissertation. Yale University, 1968.
Kent, George. "Man's Newest and Loveliest Adventure." *Reader's Digest* 64 (March 1954): 109–12.
Kirk, Andrew. "Appropriating Technology: Alternative Technology, the Whole Earth Catalog and Counterculture Environmental Politics." *Environmental History* 6 (July 2001): 374–94.
Kirsch, Scott. "Regions of Government Science: John Wesley Powell in Washington and the American West." *Endeavour* 223 (1999): 155–58.
Klein, Kerwin Lee. "Reclaiming the 'F' Word, or Being and Becoming Postwestern." *Pacific Historical Review* 65 (May 1996): 179–216.
Kroll, Gary. "The Pacific Science Board in Micronesia: Preparation and Preservation of a New Frontier Territory." *Minerva* 41, no. 1 (2003): 25–46.
———. "Rachel Carson's *Silent Spring:* Mass Media and the Origins of Modern Environmentalism." *Public Understandings of Science* 10 (2001): 403–20.

Lafollette, Marcel C. *Making Science Our Own: Public Images of Science, 1910–1955*. Chicago: University of Chicago Press, 1990.

Landauer, Lyndall Baker. "From Scoresby to Scammon: Nineteenth Century Whalers in the Foundations of Cetology." Ph.D. dissertation, 1982.

Lane, Ferdinand. *The Mysterious Sea.* Garden City, N.Y.: Doubleday, 1947.

Lear, Linda. *Rachel Carson: Witness for Nature.* New York: Henry Holt, 1997.

Lear, Linda, ed. *Lost Woods: The Discovered Writing of Rachel Carson.* Boston: Beacon Press, 1998.

Lears, T. J. Jackson. *Fables of Abundance: A Cultural History of Advertising in America.* New York: Basic Books, 1994.

———. *No Place of Grace: Antimodernism and the Transformation of American Culture, 1880–1920*. New York: Pantheon Books, 1981.

Lewenstein, Bruce V. "Public Understanding of Science in America, 1945–1965." Ph.D. dissertation. University of Pennsylvania, 1987.

Lewis, Martin, and Karen Wigen. *The Myth of Continents: A Critique of Metageography.* Berkeley: University of California Press, 1997.

Lieber, Leslie. "The World's Most Dangerous Classroom." *Science Digest,* December 1946, 84–87.

Limerick, Patricia Nelson. "Seeing and Being Seen: Tourism in the American West." In *Seeing and Being Seen: Tourism in the American West,* edited by David M. Wrobel and Patrick T. Long, 39–58. Lawrence: University Press of Kansas, 2001.

Lindgren, James. "Let Us Idealize Old Types of Manhood: The New Bedford Whaling Museum, 1903–1941." *New England Quarterly* 72, no. 2 (June 1999): 163–206.

Looker, Samuel J. Introduction to *Jefferies' England: Nature Essays by Richard Jeffries,* xi–xvii. New York: Harper and Brothers Publishers, 1938.

Lucas, Frederic. *Fifty Years of Museum Work: Autobiography, Unpublished Papers, and Bibliography of Frederic Lucas.* New York: American Museum of Natural History, 1933.

———. "The Passing of the Whale." *Scientific American Supplement* 66, no. 1717 (November 28, 1908): 337, 344–46.

———. "The Whale-Hunting Industry." *Scientific American Supplement* 65, no. 1671 (January 11, 1908): 30+.

Lundberg, Madeleine. "Eugenie Clark: Shark Tamer." *Ms. Magazine,* August 1979, 15.

Lutts, Ralph H. "Chemical Fallout: *Silent Spring,* Radioactive Fallout, and the Environmental Movement." In *And No Birds Sing: Rhetorical Analyses of Rachel Carson's "Silent Spring,"* edited by Craig Waddell, 17–41. Carbondale: Southern Illinois University Press, 2000.

———. *The Nature Fakers: Wildlife, Science and Sentiment.* Golden, Colo.: Fulcrum Publishing, 1990.

Madsen, Axel. *Cousteau: An Unauthorized Biography.* New York: Beaufort Books, 1986.

Marx, Leo. *The Machine in the Garden: Technology and the Pastoral Ideal in America.* London: Oxford University Press, 1965.

Marx, Wesley. *The Fragile Ocean.* New York: Ballantine Books, 1967.

Mathews, Eleanor. *Ambassador to the Penguins: A Naturalist's Year Aboard a Yankee Whaleship.* Boston: David R. Godine, 2003.

Matsen, Bradford. *Descent: The Heroic Discovery of the Abyss.* New York: Pantheon Books, 2005.

May, Elaine Tyler. *Homeward Bound: American Families in the Cold War Era.* New York: Harper Collins, 1988.

———. "Rosie the Riveter Gets Married." In *The War in American Culture: Society and Consciousness during World War II,* edited by Lary May, 128–43. Chicago: University of Chicago Press, 1996.

McCay, Mary A. *Rachel Carson.* New York: Twayne Publishers, 1993.

McCurdy, Howard E. *Space and the American Imagination.* Washington: Smithsonian Institution Press, 1997.

McEvoy, Arthur. *Fisherman's Problem: Ecology and Law in the California Fisheries, 1850–1980.* New York: Cambridge University Press, 1990.

McKinsey, Elizabeth. *Niagara Falls: Icon of the American Sublime.* Cambridge: Cambridge University Press, 1985.

Melville, Herman. *Moby Dick; or, The Whale.* Norwalk, Connecticut: The Easton Press, 1971.

Merriman, Daniel. "Food Shortages and the Sea." *Yale Review* 39 (March 1950): 430–44.

Mills, Eric. *Biological Oceanography: An Early History, 1870–1960.* Ithaca: Cornell University Press, 1989.

———. "The Historian of Science and Oceanography after Twenty Years." *Earth Sciences History* 12 (1993): 5–18.

Mitchell, Lee Clark. *Witnesses to a Vanishing America: The Nineteenth-Century Response.* Princeton, N.J.: Princeton University Press, 1981.

Mitchell, W. J. T. *The Last Dinosaur Book.* Chicago: University of Chicago Press, 1998.

Mitman, Gregg. *Reel Nature: America's Adventure with Wildlife on Film.* Cambridge: Harvard University Press, 1999.

———. "When Nature Is the Zoo." In *Science in the Field,* edited by Henrika Kuklick and Robert Kohler. *Osiris,* 2nd ser., 11 (1996): 117–43.

Muir, John. *The Cruise of the Corwin.* New York: Houghton Mifflin, 2000.

Mukerji, Chandra. *A Fragile Power: Scientists and the State.* Princeton, N.J.: Princeton University Press, 1989.

Murphy, Grace E. Barstow. *There's Always Adventure: The Story of a Naturalist's Wife.* New York: Harper & Brothers, 1943.

Murphy, Robert Cushman. "Animal Geography: A Review." *Geographical Review* 28 (January 1938): 140–44.

———. "Avian Orders of the Tubinares." M.S. thesis. Columbia University Press, 1918.

———. *Bird Islands of Peru: The Record of a Sojourn on the West Coast.* New York: G. P. Putnam's Sons, 1925.

———. "Conservation." *Bulletin of the Garden Club of America* (May 1937): 32–40.

———. "Conservation II." *Bulletin of the Garden Club of America* (January 1938): 39–50.

———. "Conservation and Scientific Forecast." *Science* 93 (June 27, 1941): 603–9.

———. "Conservation's Silver Lining." *Natural History* 46 (December 1940): 294–303.

———. *Fish-Shape Paumanok.* Philadelphia: American Philosophical Society, 1964.

———. "The Impact of Man upon Nature in New Zealand." *Proceedings of the American Philosophical Society* 95 (December 1951): 569–82.

———. "The Littoral of Pacific Colombia and Ecuador." *Geographical Review* 29 (January 1939): 1–33.

———. *Logbook for Grace.* New York: Time Incorporated, 1965.

———. "Lo, the Poor Whale." *Science* 91 (April 19, 1940): 373–76.

———. "The Most Valuable Bird in the World." *National Geographic Magazine* 46 (September 1924): 278–302.

———. "The Need of Insular Exploration as Illustrated by Birds." *Science* 88 (December 9, 1938): 533–39

———. "A New Zealand Expedition of the American Museum of Natural History." *Science* 108 (October 29, 1948): 463–64.

———. "The Oceanography of the Peruvian Littoral with Reference to the Abundance and Distribution of Marine Life." *Geographical Review* 13 (January 1923): 64–85.

———. "Peru Profits from Sea Fowl." *National Geographic* 115 (March 1959): 359–413.

———. "The Progress of Science." *Scientific Monthly* 56 (June 1943): 570–73.

———. "The Seventh Pacific Science Congress." *Scientific Monthly* 69 (August 1949): 84–92.

———. "South Georgia, an Outpost of the Antarctic." *National Geographic* 41 (April 1922): 408–44.

———. "The Status of Sealing in the Sub-Antarctic Atlantic." *Scientific Monthly* 7 (August 1918): 112–19.

———. "Sub-Antarctic Whaling." *Sea Power* 3 (September 1917): 44–47.

———. "The Way of the Sperm Whaler, Part I." *Sea Power* 2 (June 1917): 50–54;

———. "The Way of the Sperm Whaler, Part II." *Sea Power* 3 (July 1917): 52–57.

———. "The Way of the Sperm Whaler, Part III." *Sea Power* 3 (August 1917): 50–54.

———. "Whitney Wing." *Natural History* 44 (September 1939): 101.

Nadel, Alan. *Containment Culture: American Narratives, Postmodernism, and the Atomic Age.* Durham, N.C.: Duke University Press, 1995.

Nash, Roderick. *Wilderness and the American Mind.* New Haven, Conn.: Yale University Press, 1967.

Nelkin, Dorothy. *Selling Science: How the Press Covers Science and Technology.* New York: W. H. Freeman, 1987.

Newman, Louise. "Coming of Age, but Not in Samoa: Reflections on Margaret Mead's Legacy for Western Liberal Feminism." *American Quarterly* 48 (June 1996): 233–70.

Nichols, John T. "Life in the Bathysphere." *Saturday Review of Literature* 11 (December 8, 1934): 336.

Nicolson, Marjorie Hope. *Mountain Gloom and Mountain Glory: The Development of the Aesthetics of the Infinite.* Ithaca: Cornell University Press, 1959.

Norwood, Vera. *Made from This Earth: American Women and Nature.* Chapel Hill: University of North Carolina Press, 1993.

Novak, Barbara. *Nature and Culture: American Landscape and Painting, 1825–1875.* Oxford: Oxford University Press, 1980.

Nye, David. *America as Second Creation: Technology and Narratives of New Beginnings.* Cambridge, Mass.: MIT Press, 2003.

———. *American Technological Sublime.* Cambridge, Mass.: MIT Press, 1994.

———. *Narratives and Spaces: Technology and the Construction of American Culture.* New York: Columbia University Press, 1997.

Nyhart, Lynn K. "Natural History and the 'New' Biology." In *Cultures of Natural History,* edited by Nicholas Jardine, J. A. Secord, and E. C. Spary, 426–43. Cambridge: Cambridge University Press, 1996.

O'Brien, Raymond. *American Sublime: Landscape and Scenery of the Lower Hudson Valley.* New York: Columbia University Press, 1981.

Ocean Blueprint for the 21st Century: Final report of the U.S. Commission on Ocean Policy. 2004. http://purl.access.gpo.gov/GPO/LPS59977.

Oelschlaeger, Max. *The Idea of Wilderness.* New Haven, Conn.: Yale University Press, 1991.
Oreskes, Naomi. "Objectivity or Heroism? On the Invisibility of Women in Science." In *Science in the Field,* edited by Henrika Kuklick and Robert Kohler. *Osiris,* 2nd ser., 11 (1996): 87–113.
Our Nation and the Sea: A Plan for National Action. Report of the Commission on Marine Science, Engineering and Resources. Washington, D.C.: U.S. Government Printing Office, 1969.
Osborn, Fairfield. *Our Plundered Planet.* New York: Grosset & Dunlap, 1948.
Panetta, Leon. *America's Living Oceans: Charting a Course for Sea Change.* Summary report. Recommendations for a new ocean policy by the Pew Oceans Commission. 2003. http://www.pewoceans.org/.
Pauly, Philip. *Controlling Life: Jacques Loeb and the Engineering Ideal in Biology.* New York: Oxford University Press, 1987.
———. "Summer Resort and Scientific Discipline: Woods Hole and the Structure of American Biology, 1882–1925." In *The American Development of Biology,* edited by Ronald Rainger, Keth Benson, and Jane Maienschein, 121–50. New Brunswick, N.J.: Rutgers University Press, 1991.
———. "The World and All That Is in It: The National Geographic Society, 1888–1918." *American Quarterly* 31 (1979): 517–32.
Phinizy, Coles. "Lovely Lady with a Very Fishy Reputation." *Sports Illustrated* 23 (October 4, 1965): 50+.
Pinchot, Gifford. *To the South Seas.* Philadelphia: John C. Winston, 1930.
Potter, George Reuben. "William Beebe: His Significance to Literature." In *Essays in Criticism,* 203–28. Berkeley: University of California Press, 1929.
Pratt, Mary Louise Pratt. *Imperial Eyes: Travel Writing and Transculturation.* London: Routledge, 1992.
Price, Jennifer. *Flight Maps: Adventures with Nature in Modern America.* New York: Basic Books, 1999.
Preston, Douglas J. *Dinosaurs in the Attic: An Excursion into the American Museum of Natural History.* New York: St. Martin's Press, 1986.
Pyne, Stephen. *The Ice: A Journey to Antarctica.* Iowa City: University of Iowa Press, 1986.
Raglon, Rebecca. "Rachel Carson and Her Legacy." In *Natural Eloquence: Women Reinscribe Science,* edited by Barbara Gates and Anne Shteir, 196–214. Madison: University of Wisconsin Press, 1997.
Rainger, Ronald. *An Agenda for Antiquity: Henry Fairfield Osborn and Vertebrate Paleontology at the American Museum of Natural History, 1890–1935.* Tuscaloosa: University of Alabama Press, 1991.
Rainger, Ronald, Keith R. Benson, and Jane Maienschein, eds. *The American Development of Biology.* New Brunswick, N.J.: Rutgers University Press, 1991.
Reidy, Michael, Gary Kroll, and Erik Conway. *Exploration and Science: Social Impact and Interaction.* Santa Barbara: ABC-CLIO, 2007.
Reiger, John. *American Sportsmen and the Origins of Conservation.* New York: Winchester Press, 1975.
Revelle, Roger. "The Oceanographic and How it Grew." In *Oceanography and the Past,* edited by Mary Sears and Daniel Merriman, 10–24. New York: Springer-Verlag, 1980.

Richards, Robert. "The Structure of Narrative Explanation in History and Biology." In *Methodologies of Historical Explanations,* edited by Matthew H. Nitecki and Doris V. Nitecki, 19–53. Albany: State University of New York Press, 1992.

Richards, Thomas. *The Imperial Archive: Knowledge and the Fantasy of Empire.* New York: Verso, 1993.

Riffenburgh, Beau. *The Myth of the Explorer: The Press, Sensationalism, and Geographical Discovery.* New York: Oxford University Press, 1994.

Riley, Gordon A. "Food from the Sea." *Scientific American* 181 (October 1949): 16–19.

Roberts, Callum. *The Unnatural History of the Sea.* Washington: Island Press, 2007.

Robinson, Michael F. *The Coldest Crucible: Arctic Exploration and American Culture.* Chicago: University of Chicago Press, 2006.

Rolle, John. *Charles Fremont: Character as Destiny.* Norman: University of Oklahoma Press, 1991.

Roosevelt, Theodore. "Conservation." In *The New Nationalism,* by Theodore Roosevelt, edited by W. E. Leuchtenburgh, 50–66. New York: Prentice-Hall, 1961.

———. Review of *Jungle Peace,* by William Beebe. *New York Times Review of Books,* October 13, 1918, BR1.

Ross, Andrew. *The Chicago Gangster Theory of Life: Nature's Debt to Society.* New York: Verso, 1994.

Rossiter, Margaret. *Women Scientists in America before Affirmative Action, 1940–1972.* Baltimore: Johns Hopkins University Press, 1995.

———. *Women Scientists in America: Struggles and Strategies to 1940.* Baltimore: Johns Hopkins University Press, 1982.

Rotundo, Anthony E. "Boy Culture: Middle-Class Boyhood in Nineteenth-Century America." In *Meanings for Manhood: Constructions of Masculinity in Victorian America,* edited by Mark Carnes and Clyde Griffen, 15–36. University of Chicago Press, 1990.

Rozwadowski, Helen. *Fathoming the Ocean: Discovery and Exploration of the Deep Sea.* Cambridge: Harvard University Press, 2005.

———. *The Sea Knows No Boundaries: A Century of Marine Science under ICES.* Copenhagen: International Council for the Exploration of the Sea, 2002.

———. "Small World: Forging a Scientific Maritime Culture for Oceanography." *Isis* 86 (September 1996): 409–29.

Rubin, Joan Shelly. *The Making of Middlebrow Culture.* Chapel Hill: University of North Carolina Press, 1992.

Rydell, Robert W. *World of Fairs: The Century-of-Progress Expositions.* Chicago: University of Chicago Press, 1993.

Safina, Carl. *Song for the Blue Ocean: Encounters Along the World's Coasts and Beneath the Seas.* New York: Henry Holt, 1997.

Sapolsky, Harvey M. *Science and the Navy: The History of the Office of Naval Research.* Princeton, N.J.: Princeton University Press, 1990.

Sapp, Jan. *What is Natural? Coral Reef Crisis.* New York: Oxford University Press, 1999.

Scheese, Don. *Nature Writing: The Pastoral Impulse in America.* New York: Simon & Schuster Macmillan, 1996.

Schlee, Susan. *The Edge of an Unfamiliar World: A History of Oceanography.* New York: E. P. Dutton, 1973.

———. *On Almost Any Wind: The Saga of the Oceanographic Vessel "Atlantis."* Ithaca: Cornell University Press, 1978.

Schmitt, Peter J. *Back to Nature: The Arcadian Myth in Urban America.* New York: Oxford University Press, 1969.

Scott, James C. *Seeing Like a State: How Certain Schemes to Improve the Human Condition Have Failed.* New Haven, Conn.: Yale University Press, 1998.

Secord, Ann. "Artisan Botany." In *Cultures of Natural History,* edited by Nicholas Jardine, J. A. Secord, and E. C. Spary, 378–93. Cambridge: Cambridge University Press, 1996.

———. "Science in the Pub: Artisan Botanists in Early Nineteenth Century Lancashire." *History of Science* 32 (1994): 269–315.

Sellers, Richard. *Preserving Nature in the National Parks: A History.* New Haven, Conn.: Yale University Press, 1997.

Shershow, Harry. "Fun Under Water." *Popular Science Monthly* 148 (April 1946): 113–15.

Shoemaker, Nancy. "Whale Meat in American History." *Environmental History* 10, no. 2 (April 2005): 269–94.

Shor, Elizabeth Noble. *Scripps Institution of Oceanography: Probing the Oceans, 1936–1976.* San Diego: Tofua Press, 1978.

Shurcliff, W. A. *Bombs at Bikini: The Official Report of Operation Crossroads.* New York: W. H. Wise, 1947.

Simpson, John Warfield. *Visions of Paradise, Glimpses of Our Landscape's Legacy.* Berkeley: University of California Press, 1999.

Slotkin, Richard. *Gunfighter Nation: The Myth of the Frontier in Twentieth-Century America.* New York: Harper Perennial, 1993.

Smith, Michael L. *Pacific Visions: California Scientists and the Environment, 1850–1915.* New Haven, Conn.: Yale University Press, 1987.

Smocovitis, Vassiliki Betty. *Unifying Biology: The Evolutionary Synthesis and Evolutionary Biology.* Princeton, N.J.: Princeton University Press, 1996.

Spence, Mark. *Dispossessing the Wilderness: Indian Removal and the Making of the National Parks.* New York: Oxford University Press, 1999.

Steiner, Michael. "From Frontier to Region: FJT and the New Western History." *Pacific Historical Review* 64 (November 1996): 479–503.

Stephens, William M. "The Lady and the Sharks." *Saturday Evening Post* 232 (July 4, 1959): 52–53.

Stilgoe, John. *Alongshore.* New Haven, Conn.: Yale University Press, 1994.

———. "Bikinis, Beaches, and Bombs: Human Nature on the Sand." *Orion Nature Quarterly* 3 (1984): 4–15.

Tassen, John. "Tourists in the Underwater World." *New York Times Magazine,* June 27, 1954, 18+.

Taylor, Peter. "Technocratic Optimism, H. T. Odum and the Partial Transformation of Ecological Metaphor after World War II." *Journal of the History of Biology* 21 (1988): 213–44.

Teale, Edwin Way. *North with the Spring: A Naturalist's Record of a 17,000-Mile Journey with the North-American Spring.* New York: Dodd, Meade, 1951.

Terrall, Mary. "Heroic Narratives of Quest and Discovery." *Configurations* 6 (spring 1998): 223–42.

Thompson, Laura. "The Basic Conservation Problem." *Scientific Monthly* 68 (February 1949): 129–31.

Tinker, Spencer, and Marian Omura. *Directory of the Public Aquaria of the World.* Honolulu: University of Hawaii, 1963.

Tobey, Ronald C. *The American Ideology of National Science, 1919–1930.* Pittsburgh, Pa.: University of Pittsburgh Press, 1971.

Tomlinson, H. M. *Sea and the Jungle.* New York: E. P. Dutton, 1920.

Townsend, Kim. *Manhood at Harvard: William James and Others.* New York: W. W. Norton, 1996.

Tracey, Henry Chester. *American Naturists.* New York: E. P. Dutton, 1930

Trader Vic. *Trader Vic's Book of Food and Drink.* New York: Doubleday, 1946.

Tuan, Yi-Fu. *Space and Place: The Perspective of Experience.* Minneapolis: University of Minnesota Press, 1977.

Tucker, Jennifer. "Voyages of Discovery on Oceans of Air: Scientific Observation and the Image of Science in an Age of 'Balloonacy.'" In *Science in the Field,* edited by Henrika Kuklick and Robert Kohler. *Osiris,* 2nd ser., 11 (1996): 144–76.

Turtle, William M., Jr. *"Daddy's Gone to War": The Second World War in the Lives of America's Children.* New York: Oxford University Press, 1993.

Van Dervoort, J. W. *The Water World, or The Ocean, Its Laws, Currents, Tides, Wind-Waves, Phenomena, Mechanical Appliances, Animal and Vegetable Life.* New York: Union Publishing House, 1883.

Verge, Arthur C. "George Freeth: King of the Surfers and California's Forgotten Hero." *California History* 80 (2001): 82–105.

Vogt, William. *Road to Survival.* New York: William Sloane Associates, 1948.

Walton, Izaak. *The Complete Angler, or The Contemplative Man's Recreation.* Boston: Little, Brown, 1912.

Wandersee, Winifred D. *Woman's Work and Family Values, 1920–1940.* Cambridge, Mass.: Harvard University Press, 1981.

Warren, Louis. *The Hunter's Game: Poachers and Conservationists in Twentieth-Century America.* New Haven, Conn.: Yale University Press, 1997.

Welker, Robert Henry. *Natural Man: The Life of William Beebe.* Bloomington: Indiana University Press, 1975.

Weisgall, Johnathan M. *Operation Crossroads: The Atomic Tests at Bikini Atoll.* Annapolis: Naval Institute Press, 1994.

White, G. Edward. *The Eastern Establishment and the Western Experience: The West of Frederic Remington, Theodore Roosevelt, and Owen Wister.* New Haven, Conn.: Yale University Press, 1968.

White, Richard. *The Organic Machine.* New York: Hill and Wang, 1995.

White, Richard, Patricia Nelson Limerick, and James Grossman. *The Frontier in American Culture: An Exhibition at the Newberry Library, August 26, 1944–January 7, 1955.* Berkeley: University of California Press, 1994.

Whitehead, John S. "Noncontinguous Wests: Alaska and Hawai'i." In *Many Wests: Place, Culture and Regional Identity,* edited by David M. Wrobel and Michael C. Steiner, 314–41. Lawrence: University Press of Kansas, 1997.

Willard, Michael Nevin. "Duke Kahanamokiu's Body: Biography of Hawai'i." In *Sports Matters: Race, Recreation, and Culture,* edited by John Bloom and Michael Nevin Willard, 13–38. New York: New York University Press, 2002.

Wilson, Alexander. *The Culture of Nature: North American Landscape from Disney to the Exxon Valdez.* Cambridge, Mass.: Blackwell Publishers, 1992.

Wilson, Rob. *American Sublime: The Geneaology of a Poetic Genre.* Madison: University of Wisconsin Press, 1991.

———. *Reimagining the American Pacific: From South Pacific to Bamboo Ridge and Beyond.* Durham, N.C.: Duke University Press, 2000.

Worster, Donald. "Landscape with Hero: John Wesley Powell and the Colorado Plateau." *Southern California Quarterly* 29 (1997).

———. *Nature's Economy: A History of Ecological Ideas.* Cambridge: Cambridge University Press, 1985.

Woodward, Colin. *Ocean's End: Travels through Endangered Seas.* New York: Basic Books, 2000.

Wrobel, David. "Beyond the Frontier-Region Dichotomy." *Pacific Historical Review* 65 (August 1996): 401–29.

———. *The End of American Exceptionalism: Frontier Anxiety from the Old West to the New Deal.* Lawrence: University Press of Kansas, 1993.

———. "Introduction: Tourists, Tourism and the Toured Upon." In *Seeing and Being Seen: Tourism in the American West,* edited by David M. Wrobel and Patrick T. Long, 1–34. Lawrence: University Press of Kansas, 2001.

Index

Allen, Irwin, 121
Allen, Joel, 24–25
Alvin, 182
American Geographical Society, 54
American Museum of Natural History (AMNH), 4–5, 11
  Arcturus Exhibit, 48
  conservation luncheon, 31–33
  Coral Reef Group, 48
  educational television series, *Adventure*, 137
  Fishes of the World exhibit, 29–30
  Hall of Africa, 57
  Hall of Biology of Mammals, 12
  Hall of Ocean Life, 36, 48–49, 92
  Lerner Laboratory, 131
  Roosevelt Memorial Hall, 57
  Whitney Hall of Pacific Birds, 52, 57–59
American Ornithologists Union, 23
Andrews, Roy Chapman, 3–4, 125, 192
  as administrator, 35–36
  on adventure, 17–19
  Central Asiatic Expedition, 35
  compared to Robert Cushman Murphy, 39, 42
  on conservation, 20
  controversy with whaling company, 17–18
  early history of, 11–12, 19
  efforts to encourage whale consumption, 31–33
  fieldwork in Alaska, 16
  fieldwork in Japan and Korea, 9, 23–25, 33–35
  fieldwork in Vancouver, 12–16
  fieldwork on Long Island, NY, 12
  and gender stereotypes, 127
  hunting with camera, 15
  on *Moby Dick*, 19–20
  on modern natural history, 35
  Monograph on Gray Whales, 35
  Whale Hunting with Camera and Gun, 16
  on whale meat, 30–33
  on whale overfishing, 16–17
angling, 26–27
Anthony, Harold, 95
aquariums, 91, 129, 130, 143
Aronson, Lester, 142

Ballen, Francisco, 52, 57
Barton, Otis, 65, 78, 84, 87, 91–92
  and *Titans of the Deep*, 92
Bates, Marson, 39
bathysphere, 5
  construction of, 78
  and contour dives, 85
  criticisms of, 79, 82, 86
  and film, 91–92
  and the NBC broadcast, 65–67
  at the 1930 World's Fair, 92–93
  preparation for dives, 78
  problems with acquiring specimens, 86
  scientific experiments with, 85
  seen as a stunt, 67, 84–85
Beach Boys, 166
Beebe, William, 5, 192
  Arcturus Adventure, 74–76
  Arcturus Exhibit at AMNH, 48
  Beebe Project, 124
Bermuda Oceanographic Expedition, 65–67, 76–84
  characterization of deep sea world, 73, 88

Beebe, William, *continued*
compared to Roy Chapman Andrews and Robert Cushman Murphy, 66
as conservationist, 93
criticisms of, 74–76
disassociation from bathysphere, 90–94
distaste for publicity, 66
and diving helmets, 73–74, 77
early history, 68
fieldwork in Bermuda, 76–84
fieldwork in Haiti, 76
fieldwork on pheasants, 68
and gender, 72
Half Mile Down, 67, 86
and heroism, 79–80, 90
and Jacques Cousteau, 169
"A Jungle Laboratory," 69
Jungle Peace, 69
managing sensationalism, 65–68, 70–71, 82, 83, 86, 90, 93
as nature writer, 69
NBC broadcast, 65–67
on ocean as food resource, 120
and Rachel Carson, 99, 101, 105, 110, 111, 112, 117
scientific legitimization of bathysphere, 85–89
self-effacement, 66–67, 74, 83, 84, 88–90, 94
on sharks, 145–146
station management, 77–78
in the tropics, 68–69
use of the sublime, 68, 74, 87–90
Benchley, Peter. See *Jaws*
Benson, John, 48
Bermuda Oceanographic Expedition, 65–67, 76–84
Beebe's management of, 77–78, 82
modification to Nonsuch Laboratory, 76–77
1930 season, 78–80
1932 season, 80–82
1934 season, 82–84
patronage from Bermudan government, 76–77
Bergeron, Victor, 165
Betson, Henry, 103, 105
Bigelow, Henry, 115
biocentrism, 5, 74, 93–94, 99, 102–103, 119–120, 123
biogeography, 4, 52, 54
Bond, Ford, 65–67
Bond, George, 179–181
Boone, Daniel, 176
Boone and Crocket Club, 4, 23
Bowdoin, George S., 12
Bowman, Isaiah, 54
Bradley, David, 136
Brady Bunch, The, 167
Breeder, Charles, 129, 140
Bronx Zoo. *See* New York Zoological Society
Brooks, Charles, 115
Bumpus, Herman, 17

Cannon, Berry, 181
Cape Haze Laboratory (Mote Marine Laboratory), 6, 139–147
Carpenter, Scott, 180
Carson, Rachel, 5–6, 192
and the *Baltimore Sun*, 108–109
"Conservation in Action," 107–108
conservation writing for U.S. Fish and Wildlife Service, 107–108
early history of, 100
Edge of the Sea, 123
editing work for federal agencies, 107
on experimental biology, 99–101
merging nature writing with science writing, 110–112, 114
and nature writing, 98, 102–106
and oceancentrism, 97
personal exploration of the sea, 102, 123
reviews of *Under the Sea-Wind*, 104
and Robert Cushman Murphy, 115, 117
The Sea Around Us, 125, 135, 162
and science writing, 106–111
Silent Spring, 63, 97, 117, 122, 123
and Thor Heyerdahl, 115, 162
"Undersea," 101–102
Under the Sea-Wind, 102, 106
and U.S. Bureau of Fisheries/Fish and Wildlife Service, 98, 104–105, 106–107, 110, 115–116, 120
and William Beebe, 99, 101, 111, 117
Central Asiatic Expedition, 9–10, 35
Clark, Eugenie, 6, 193
Beebe Project, 124
biographical treatments of, 125–127, 142, 148–149
demystifying the myth of dangerous sharks, 135, 139, 143, 147
and the domestication of the ocean, 125–127, 142, 146

domestic ideology of, 125–128, 130, 147–149
early life of, 128–129
family life of, 126
fieldwork in Micronesia, 131–134, 138
fieldwork in Red Sea, 134
fish as pets and research organisms, 129, 130, 132–133, 142
hermaphroditic grouper, 139–140
and Ilias Konstanitu, 126, 134, 148
Japanese ancestry and national security, 129
on *Jaws*, 145
Lady with a Spear, 6, 128, 133, 135, 148
Plectognathi research at American Museum of Natural History, 129
research at Cape Haze Laboratory, 139–147
reviews of *Lady with a Spear*, 135, 138
and Robert Cushman Murphy, 131
and second wave feminism, 149–151
and Siakong, 133–134, 138
and skin-diving, 132
and transgressing gender stereotypes, 126–128, 147–151
and the U.S. military, 131–132
and work during World War II, 129
Clark, James, 12, 25
Cleveland, Benjamin D., 42
Coker, Robert Ervin, 97, 111
cold war, 153
Conrad, Joseph, 50
conservation
of guano oceanic birds, 55–56
international conservation, 21–23, 46
federal, 98, 104, 106–110
of fisheries, 106–110
and the Maori, 62
of Pacific islands, 57–62, 117, 155
Progressive, 3–5, 11, 20, 22–23, 33
of whales, 20–23, 46–47
Cousteau, Jacques, 6–7, 137, 193
Alcyone, 188
and appropriate technology, 170, 184–188
and *Calypso*, 169, 171–172, 177, 183, 187
compared to U.S. ocean explorers, 182
compared to William Beebe and Thor Heyerdahl, 169, 170, 171, 186
Conshelf I, 175
Conshelf II, 175–176
Conshelf III, 176–177
"Cousteau Camera," 171
criticisms of, 184
domestic ideology, 176
early history of, 170
ecotechnie, 188
on egocentrism, 186
and Eugenie Clark, 183
Homo aquaticus, 174–175
invention of aqualung, 170
on limits of technology, 177
and modern environmentalism, 184
nature adventure, 183
Oasis in Space, 187
"The Oceans are Dying," 185
as ocean technician, 153, 170
The Silent World, 173
The Undersea World of Jacques Cousteau, 183–185, 187
and underwater archeology, 171
and underwater film, 170–171
on weather production centers, 186–187
work for petroleum industry, 172–173, 177
World without Sun, 176
Cousteau, Phillipe, 177
Cousteau Society, 187
Cook, Frederick, 68
Cook, James, 45, 80
Coolidge, Harold, 131
Cooper, Isabel, 71
Cronon, William, wilderness thesis, 193–194

Daisy, 38, 41–45
as an anachronism, 47–48
Darwin, Charles, 38, 47, 70, 97, 112, 117
Davenport, Charles, 48
deep sea fishes
amphioxus, 73
cyclothones, 73, 85
Diaboldium arcturi Beebe, 74
hatchet fish, 73
mychtophids, 82
pharynx fish, 73
pteropods, 82
siphonophores, 82
Sternoptyx, 73
Deep Sea System Research Group, 179
Deep Submergence Systems Review Group, 181
Department of Tropical Research, 65–67
Arcturus Expedition, 70–76
Beebe's management of, 69–70
in British Guyana, 68–69
exhibits at 1939 World's Fair, 92–93

Department of Tropical Research, *continued*
financing and patronage, 66, 69–70, 83
patronage of, 66
See also Bermuda Oceanographic Expedition
diving, helmet, 73–74, 77, 135
diving, scuba, 6, 137–138, 140, 170
compared to helmet diving, 173
diving, skin, 126, 132, 136–137
domestic ideology, 128. *See also* Clark, Eugenie; explorers; and natural history
Dos Passos, John, 154
Douglas, William O., 189
Dr. Strangelove, 145
Dreyfuss, Richard, 145
Dumas, Frederic, 170

Earle, Sylvia, 1, 144, 189
Earth Day, 185, 189
eastern establishment, 4, 10–11
Edgerton, Harold, 172
Eisenhower, Dwight, 178
Emerson, Ralph Waldo, 2, 67
environmentalism, modern, 120, 122, 168, 177
evolution of ocean life, 34–35, 43, 51
expansionism
into Pacific, 59–59, 154–155
United States, 2–3, 11
explorers
in the American West, 1–3, 39, 55
cultural work of, 1–8, 193
and gender stereotypes, 127
"going-where-no-one-has gone," 67
and heroism, 79–80
and technology, 6, 80
Explorers Club, 11, 32, 127, 158

Falla, Robert A., 59, 61
Foyn, Svend, 14–15
Freeman, Dorothy, 116
Freud, Sigmund, 87, 97
frontier
idea of, 1–2, 7–8, 10–11, 29, 33, 36, 39, 55, 125, 152–155, 166, 191
myth of the inexhaustible, 4, 22–23, 40, 46, 64
frontier anxiety, 10–11, 30

Gagnan, Emile, 170
Galápagos Islands, 70, 73–74
game fish
bonefish, 28
broadbill swordfish, 27–28
marlin, 27–28
sailfish, 28
striped bass, 28
tarpon, 27–28
yellowfin tuna, 27
Garden Club of America, 40
Gilbert, Perry, 143
Grant, Madison, 72, 76, 83
Gregory, William, 29
Grey, Zane
and deep sea fishing, 28–30
Tales of Fishing Virgin Seas, 29
trophies on display at the AMNH, 29–30
guano
Administration of Peru, 52, 55–56
conservation of, 55–56
islands of Peru, 4, 51–56

Haas, Hans, 136, 138
Halle, Louis, 111–112
Heller, John, 140, 142
Hemingway, Ernest, 29, 114
hermaphroditism, in grouper, 139–140
heroism. *See* explorers; natural history
Heyerdahl, Thor
and academic anthropology, 158
advertisements for *Kon-Tiki*, 163
American Indians in the Pacific, 163
Association of World Federalists, 168
back to nature, 157, 161, 162
compared to Rachel Carson, 115, 156–157, 162, 169
critique of technology, 152–153
early history of, 156–157
and the Explorer's Club, 158
Fatu-Hiva, 157
fieldwork in Easter Island, 167–168
fieldwork in Fatu Hiva, 157–158
film adaptation of *Kon-Tiki*, 159
Kon-Tiki: Across the Pacific in a Raft, 152–153, 154
observing ocean pollution, 168
Ra expeditions, 168
reviews of *Kon-Tiki*, 163–164
South American Indian migration theory, 153, 158
United Nations Conference on the Human Environment, 168
See also *Kon-Tiki* expedition; Kon-Tiki fever

Higgins, Elmer, 101
Hollister, Gloria, 78
Homestead Act, 55
Hornaday, William, 31
Howe, Quincy, 102
Hubbs, Carl, 86, 130–131
Humboldt Current, 51
  and the *Arcturus* Expedition, 70, 72
  ecology of the, 52–53
  historical ecology of the, 53–54
  and *Kon-Tiki*, 160
hunting
  with camera, 9, 16
  and the eastern establishment, 10–11
  and masculinity, 10, 19, 136
  ocean fauna, 26–30
  spearfishing, 136
  as sport, 4
  whales, 9

imperialism, culture of, 23–24, 60–61, 131–132, 134
International Council for the Exploration of the Seas, 21
International Underwater Spearfishing Association, 136
International Whaling Commission, 46
Irving, Washington, 2

Japanese whale fishery, 30–31
Jaws, 144–145, 146
Jeffries, Richard, 103
Jessup, Morris K., 12
Johnson, Lyndon B., 178, 182
Jones, James, 138
Jordan, David Starr, 21

Kahanamoku, Duke, 166
Kaplan, Moise N., 28
Keats, John, 87
Kennan, George, on containment, 153
Kennedy, John, 178
King, Clarence, 2, 5
Kinsey, Alfred, 128, 130
Kipling, Rudyard, 88
Kon-Tiki expedition
  as anodyne to cold war politics, 162
  building raft with natural material, 159
  as critique of modern science and technology, 163–164
  emergency equipment of, 159
  juxtaposed with modern navies, 159
  observing and eating sea-life, 161
  origins of, 157–158
  voyage, 160–162
Kon-Tiki fever, 6, 154, 164–167
  as anodyne for cold war anxiety, 167
  and "The Brady Bunch," 167
  consumerism and postwar culture of affluence, 165–167
  Disneyland's Tiki Room, 165
  middle-class Pacific vacations, 165–166
  Polynesia-themed restaurants and hotels, 165
  surfing, 166

Lane, Ferdinand, 111
Leopold, Aldo, 1, 123
life zones, 54–55
Light Tackle Club, 28
Long, John, 83–85
Long Island Conservation Association, 63–64
Long Key Fishing Camp, 28
Lorenz, Pare, 121
Lucas Frederick
  on classic versus modern whaling, 22–23, 43
  disagreement with J. A. Morch, 42
  and Robert Cushman Murphy, 41
  on whale meat, 31

Man in the Sea Program (U.S. Navy), 179–181
  and publicity, 180
  testing human extremes, 180
Man Invades the Sea, 180
Marine Protected Areas, 191
Marine Protection, Research and Sanctuaries Act, 190
Marsh, George Perkins, *Man and Nature*, 40, 46
Marx, Wesley, *The Frail Ocean*, 189–190
Maury, Matthew Fontaine, 55
Melville, Herman, 2, 19–20, 47, 145
Merriam, C. Hart, 55
Merriman, Daniel, 115, 120
Miami Beach Rod and Reel Club, 28
Michener, James, 163–164
Micronesia, 6, 117
  as prelapsarian paradise, 155–156
  Scientific Investigations of Micronesia, 131–134
  U.S. Trust Territory, 154

middlebrow culture, 102–103
Millikan, Robert, 85
Morgan, J. P., 12
Muir, John, 2, 5, 67, 112, 123, 170
Murphy, Grace Barstow, 41, 63
Murphy, Robert Cushman, 4–5, 71, 192
  aboard a steam whaler, 45–46
  and AMNH, 4
  Bird Islands of Peru, 52
  as classic naturalist, 39
  on conservation, 39–40
  on conservation of Pacific islands, 57–62
  critique of natural resource use, 40
  A Dead Whale, or a Stove Boat, 47
  early history of, 41
  environmental history of Long Island, NY, 63–64
  and Eugenie Clark, 131
  on evolution of Tubinares, 51
  fieldwork in Columbian Bight, 56–57
  fieldwork in the guano islands, 51–56
  fieldwork in Long Island, NY, 63
  fieldwork in New Zealand, 59–61
  fieldwork in the sub-Antarctic, 41–47
  Fish-Shape Paumanok, 62–63
  as geographer, 53–54
  as historian of sperm fishery, 47–51
  "Impact of Man upon Nature in New Zealand," 61–62
  investigation of ocean birds, 43
  Logbook for Grace, 47
  on modern whaling, 46
  Nantucket sleigh ride, 37–38
  on ocean biogeography, 54–55
  Oceanic Birds of South America, 54
  oceanography, 53–54
  and Rachel Carson, 115, 117
  and whaleboat exhibit, 48–49

National Geographic Society
  Beebe Project, 124
  and the Bermuda Oceanographic Expedition, 82–84
  and Eugenie Clark, 144–145
  and Jacques Cousteau, 170, 172, 175
  and the Sustainable Seas Project, 1
  and Sylvia Earle, 189
National Oceanic and Atmospheric Administration, 185
natural history, 4–5
  and adventure, 18–19, 71
  and bathysphere, 86
  classic, 4–5
  and conservationist activism, 64
  early-twentieth century challenges, 39–40
  and ecology, 39
  fieldwork as rite of passage, 43
  and gender stereotypes, 6, 72, 127
  and heroism, 67–68, 79–80
  and hunting, 4, 25–26
  and life histories or organisms, 43
  and nature writing, 97–98, 99
  and patronage, 77
  problems with nineteenth-century cetaceans, 33–34
  and Romanticism, 39
  and Sealab, 180
  and sensationalism and publicity, 5, 18–19, 66–67, 71, 93
  and whaling shore stations, 16, 33–35
nature writing, 67–68, 97–98
  as distinct from science writing, 109–110
  oceanic, 99
  and Rachel Carson, 102–106, 111–113
NBC broadcast, 65–67, 80–82
neo-Malthusianism, 120
New York World's Fair, 92–93
New York Zoological Society, 65–67. *See also* Department of Tropical Research
New Zealand
  impact of western colonization, 61–62
  and the Maori, 60
  moa on display at AMNH, 57
  Pelorus Jack, 60
  Pyramid Valley Swamp, 60–61
  Snares Islands, 60–61
  Waipoua Forest, 60
Niño, El, 53–54, 56, 72

ocean
  as cold war geography, 153, 178–183
  conquest/domination of, 125, 143, 176, 178
  conservation of, 40
  destruction of, 190
  domestication of the, 6, 125–127, 142, 146, 151
  ethic, 1, 153, 188
  as food resource for global hunger, 120–122
  modern crisis, 191–194
  and neo-malthusianism, 6
  nineteenth-century conceptions of, 2
  Pacific Ocean in western culture, 154

pollution, 7, 122, 168, 184
in post World War II American literature, 96
psychology of sea travel, 50
as sublime spectacle, 67–68, 94
territorialization of the ocean, 1–8, 40, 54, 191–192
as wilderness, 1, 70
ocean birds
albatross, 43, 44, 50–51
biogeography of, 54–55
cape pigeons, 51
ecology of, 50–51
and film, 91–92
giant fulmars, 51
petrels, 43, 50–51
shearwaters, 50
terns, 3
whalebirds, 51
oceancentrism, 5. *See also* Carson, Rachel: *The Sea Around Us*
oceanography
equipment, 53, 72–73, 77
problems unique to women in the field, 149–151
and *The Sea Around Us*, 114–116, 118
and World War II, 98
oceanology, 182
Oriental Whaling Company, 9, 24, 30
Operation Crossroads, 155–156
Osborn, Fairfield, 120
Osborn, Henry Fairfield, 35, 68
on bathysphere's usefulness for science, 79
on ocean environments, 51
theory of radiational adaptation, 34
thoughts on the *Arcturus* Expedition, 70
on whale meat, 31–32

Pacific Ocean. *See* Micronesia
Pacific Science Board, 61, 131
Pacific Science Congress, 61
Pacific War Memorial, 134
Pacific Whaling Company, 14–17, 18
Panetta, Leon, 1
Peary, Robert, 31–32, 68
Peru. *See* islands of Peru: guano
petroleum exploration, United States, 179
Piccard, August, 80, 92
Powell, John Wesley, 55
Putnam, George Palmer, 70–72, 80
radio. *See* NBC broadcast
Riley, Gordon, 120
Ritter, William, 108
Rodell, Marie, 116
Roman, Earl, 28
Romanticism, 39
Roosevelt, Theodore, 3, 11, 19, 57, 68, 69
Rose, Ruth, 72

Sailfish Club of Palm Beach, 28
Salt Water Anglers of America, 28
salt water game fishing, 26–30
Sanford Whitney South Pacific Expedition, 54
Sargasso Sea, 70, 71
Science Service, 108
science writing, 106–111
Scripps, Edwin, 108
Scripps Oceanographic Institute, 149
scuba. *See* diving, scuba
The Sea Around Us, 125, 135, 162
Burroughs Medal, 122
defined, 97
film adaption by RKO, 121
final effect of, 99
library research for, 99
merging nature writing with science writing, 114, 116, 117–119
National Book Award, 122–123
nature writing influences on, 111–112
1961 preface, 121
"The Ocean and a Hungry World," 121
and oceancentrism, 97, 112–118
and oceanic natural history, 114
and oceanography, 114–116, 118
origin of, 99, 111
and Pacific Islands, 116–117
in post-World War II popular culture, 97
reviews of, 113–114, 118–119
and World War II, 111–112
sea elephant hunting, 45
Sealab, 179–181
Seawolf, 178
Sedgwick, Ellery, 101
sharks
attacks on humans, 143–144
behavior research, 124, 142–143
demystifying the myth of dangerous sharks, 135, 139, 143, 144, 145
deterrents, 143–144, 159

sharks, *continued*
during World War II, 143
Jaws, 144–145
lemon, 143
liver as anti-carcinogen, 140
nurse, 6
requiem, 144
Sharks (Cousteau program), 183
Shark Research Panel, 144
tiger, 74
and William Beebe, 74
and whales, 47
whale shark, 146
white-tipped reef shark, 171
Sierra Club, 23
Skinker, Mary Scott, 98
Slocum, Kenneth, 182
Society of Women Geographers, 127
South Georgia, 4, 44–47
Spencer, Herbert, 90
stratosphere balloon flights, 83–84, 85
Stratton Commission, 190
Sub-Marine Alpine Club, 137
sublime, aesthetic of, 5, 67–68, 86–87. *See also* uses of the sublime
submersibles, 181–182
surfing, 166
Sutton, George Miksch, 105
Sverdrup, Harold, 149–150
swordfish fishery, 145

Taylor, Frederick Winslow, 15, 34
Teal, Edwin, 112
technology
appropriate, 170, 184–188
and frontiers, 152–153
Tee-Van, Helen, 71
Tee-Van, John, 78, 84
Tektite 2, 189–190
Thompson, Laura, 155
Thoreau, Henry David, 67, 98, 103, 112, 123
Townsend, Charles, 31, 48, 91
Trader Vic's, 165
Trotter, Elizabeth, 71
True, Frederick, 34–35
Tomlinson, Henry, 103
Torp, Liv Coucheron, 157
Tuna Club of Catalina, 28
Turner, Frederick Jackson, 3
Tyee Club of British Columbia, 28

United Nations Conference on the Human Environment, 168
United Nations Conference on the Law of the Sea, 187
U.S. Bureau of Fisheries, 21, 24, 32
Albatross, 23–24, 48, 106–107
and Rachel Carson, 98, 104, 106–107
uses of sublime
and deep-sea pressures, 87–88
early characterizations of the ocean sublime, 87
U.S. Fish and Wildlife Service, 5
and Eugenie Clark, 129
and Rachel Carson, 98, 121
U.S. Underwater Demolition Team, 135

Van Dervoort, J. W., *Water World*, 87
Van Loon, Henrik, 102–103
Vogt, William, 120
Voyage to the Bottom of the Sea, 178

W. R. Grace Co., 52
Walton, Isaac, *The Compleat Angler, or The Contemplative Man's Recreation*, 26
Watson, Mark, 108
West. *See* exploration; frontier
whales
blue, 15, 16, 25
commercial extinction, 17–23
declining fisheries, 4, 15, 22–23, 42
on display at the AMNH, 12, 25
finback, 14
as food, 30–33
as game, 10, 23–30
gray, 33, 35, 184
humpback, 14, 16
killer, 25, 33
as meat, 31–33
rorqual, 14
sei, 25, 35
shore station whaling, 12–23, 25, 33–35, 45–46
sperm, 15, 16, 25, 145
sperm fishery, 22, 37–39, 47–51
stranding and rescue, 95–99
whaling and cultural identity, 47–51
See also under conservation
Whitney, Caspar, 31
Whitney South Sea Expedition, 59
Whole Earth Catalog, 186

wilderness, idea of. *See* frontier
Williams, Harrison, 70, 77
Williamson, Henry, 104
Wilson, Woodrow, 31, 32
women
in post World War II science work, 127, 129–130
in post World War II workforce, 127
Woods Hole, Marine Biological Laboratory, 98, 99, 131
World Congress on Underwater Activities, 173
World War I, 31–32
and the vegetarian/feminist critique, 32
World War II
environmental impact on Pacific islands, 56–59, 154–156
and oceanography, 98
and *The Sea Around Us*, 111–112